NTUMBA BUILELA Hubert

Management's contribution to the creation of a micro-hydropower plant

NTUMBA BUILELA Hubert

Management's contribution to the creation of a micro-hydropower plant

Towards the use of renewable energy for the development of rural and urban-rural areas in the DRC

ScienciaScripts

Imprint

Cover image: www.ingimage.com

This book is a translation from the original published under ISBN 978-620-6-71516-0.

Publisher:
Sciencia Scripts
is a trademark of
Dodo Books Indian Ocean Ltd. and OmniScriptum S.R.L publishing group

120 High Road, East Finchley, London, N2 9ED, United Kingdom
Str. Armeneasca 28/1, office 1, Chisinau MD-2012, Republic of Moldova, Europe
Printed at: see last page
ISBN: 978-620-8-08899-6

Contents

Epigraph

Starting from nothing, to build a monument.

NSAMAN - O - LUTU Oscar (Ph.D)
Professor Emerite

In memoriam

To my dear father **NTUMBA BULELA Hubert**
To my dear mother **KAMWANYA Therese**

NTUMBA BUILELA Hubert

Dedication

To my dear wife and life companion AMBA BANGA Lucie

ACKNOWLEDGEMENTS

As a preamble to this memoir, I would first of all like to thank my Lord Jesus Christ who gave me the potential to complete this work and I would like to express my most sincere thanks for the breath of life and his grace and mercy.

I would like to express my sincere thanks to Professor PALAMA BONGO Francois, who did me the honour of being the Director of this work, as well as for his supervision, his many pieces of advice and his constant support throughout this subject.

I would like to sincerely thank the co-directors Professor MBOL Bernadin, Professor MUAKA NDOMBE Justin and Professor MANY Emmanuel who helped me and contributed to the drafting of this dissertation, for their presence and for the time they were willing to devote to re-evaluating this work, and the members of the postgraduate coordination team at CEPROMAD University.

My thanks also go to Professor MAVUELA Richard for all she has done for such a just and necessary cause, for being able to encourage me all the time in my noble decision.

I would like to extend a very special and religious thank you to Professor Emerite NSAMAN - O - LUTU Oscar (Ph.D), initiator, magnificent Rector, Professor Emerite and the University Professors of the CEPROMAD network.

My heartfelt thanks go to Professors ATWEL - OKEL MUNTUNGI Gode and MWAMBA Gode for their merit in encouraging me and pushing me to go further than limiting myself to the lower level after obtaining my bachelor's degree.

I would like to express my sincere thanks to the Professors and administrative staff of CEPROMAD University, and to the Professors of other universities for their willingness and availability to help me carry out and complete this research.

Finally, I would like to express my sincerest thanks to the people who have always supported and encouraged me during the writing of this thesis.

NTUMBA BULELA Hubert

GENERAL INTRODUCTION

1. CONTEXT AND FRAMEWORK OF THE WORK[1]

This dissertation is designed to be defended in order to obtain the Diploma of Advanced Studies in Management and Economics, majoring in Business Management. As part of the organisation of the third cycle at the University of CEPROMAD, UNIC in short, which recommends that DEA students write and defend in public a dissertation on the research project, the results of which will be of social, economic and environmental utility, with objectively verifiable indicators, in order to sanction the end of the 3rd cycle by implementing a micro-hydropower plant at Kinzono, together with a dose of management to promote development.

At the thirty-fourth meeting of the Council of Ministers held in Lubumbashi on 24 December 2021, the Congolese government approved the local development programme for 145 territories (PDL - 145 T).

The contribution of both private and public development partners (investors) will also be sought to speed up its implementation for the benefit of the local population. The main aims of this development are to reduce social inequalities, revitalise local economies and transform the living conditions of the Congolese population living in areas that have hitherto been poorly served by the planning and programming system, as part of efforts to combat the territorial inequalities, insecurity and multidimensional poverty facing rural populations.

This ambitious model is divided into two components:

1. Improving access to electrification and basic socio-economic services for people living in rural areas.
2. Promote the development of rural economies and local value chains by combating poverty.

Our work is entitled "Management de creativite et strategie de développement de projet de la microcentrale de Kinzono, ville province de Kinshasa dans la commune de Maluku. An analytical study of the data

[1] A. DAYAN et al, Manuel de gestion, volume 1,2e ed Elipses, Paris, 2004, p43

Quantitative and qualitative studies are still essential to unravel, step by step, the phenomena that accompany the "creativity and strategic management" project or model in the city of Kinshasa, specifically in the commune of Maluku, in the KINZONO district, which has an urban-rural character (a constraint on the city), identified by underdevelopment and the absence of electricity.

These to identify the factors that are adversely affecting the area's progress. By carrying out this study to investigate the feasibility of this development, taking into account the managerial and strategic aspects of the site.

This implementation and the implementation of this activity, encouraged us to manage in a assiduous way the managerial cultures. This scientific reflection is divided into two main parts, apart from the general introduction and the general conclusion. The first is the conceptual study, subdivided into two chapters, namely generalities on general operational and strategic management, followed by the notion of creation, which draws on knowledge of creativity management, the strengths of Michael Porter, and standards (ISO). Performance management models and management styles, without forgetting the concepts of BLAKE and MOUTON, PESTEL, MCKINSEY, etc.

Part Two, entitled Economic Management Model, is also subdivided into two chapters: the first deals with the concept of energy, which leads to the efficient use of energy, and the second deals with energy management, environmental studies, the impact of energy (development) and finally the proposed model resulting from the study.

2. CHOICE AND INTEREST OF SUBJECT

First and foremost, you need to know that no activity can develop without energy supply and distribution to the population, which means that we are obliged to map the production site and take stock of the situation.

Given that energy is an essential factor in the development of all areas of life. Energy is defined as a people's ability to harness nature, transforming it and using it for their own well-being. Acting as an agent for the development of the area, electrical energy will support the development of this district, by setting up this micro-hydro plant that will contribute to the local economy of the inhabitants. The aim is to provide a hypothetical basis for more in-depth studies by ourselves or by other researchers interested in the field. This work covers several areas of interest, namely:

a) Scientific interest: This study looks into the causes of the lack of electrification, with a view to formulating hypotheses for adjusting the methodological approaches that could lead to peaceful development.

b) Personal interest: The installation of radios and televisions has led to the development of information, a factor that is conducive to culture.

c) Collective interest: The demonstration of lighting would promote security and peace in the national interest.
d) National interest: The micropower station will contribute to development where energy is available.
e) Economic interest: Electricity users pay taxes in order to increase the public treasury and meet the country's many needs.
f) Ecological interest: The use of electrical energy helps to reduce excessive consumption of wood;
g) Health benefits: Electricity makes it easier for REGIDESO to distribute water, which reduces diseases caused by dirty hands and avoids the death of weeds.
The Democratic Republic of Congo suffers from a shortage of electricity, particularly in urban and rural areas such as the KINZONO district in the commune of MALUKU. Studies have shown that the electrification rate in DR Congo is 31%, with 30% in urban areas and 1% in urban-rural areas. In Europe, 95% of the population work in the formal sector, while in Africa the same percentage work in the informal sector due to underdevelopment. The main reason for this is the lack of electricity in neighbourhoods, which results in disastrous living conditions. What are the causes of underdevelopment in urban and rural areas? And what mechanism can be put in place to solve the problem of under-equipment?
This study will help to compare ideas and clarify the opinions of stakeholders who are capable of creating the right development climate so that communities can benefit from appropriate services and be managed according to the principles of good local governance.
The primary aim of this scientific reflection is to make an analytical and managerial contribution to reducing the rate of underdevelopment in relation to the Millennium Development Goals (MDGs).
The eight Millennium Development Goals (MDGs) were adopted in New York in 2000 as part of the United Nations Millennium Declaration by 193 UN member states and at least 23 international organisations, which agreed to achieve them by 2015. The Sustainable Development Goals (SDGs), which follow on from the MDGs, were then drawn up. These goals cover major issues and are structured around these objectives.
Reducing extreme poverty and each of the development goals can be broken down into several targets.
This objective is based on three targets.
❖ **The first target: to halve, between 1990 and 2015, the proportion of people living on less than a dollar a day.**

The World Bank estimates that in 2005, 1.4 billion people were living in extreme poverty. The food crisis, caused by rising commodity prices, is pushing around 100 million more people into extreme poverty. If this target seems within reach, it is mainly because of economic growth in Asia, while Sub-Saharan Africa appears to be stagnating.

❖ **Target 2: Provide decent, productive employment for the entire population.**

Over the last ten years, productivity in Asian and CIS countries has quadrupled, helping to reduce the number of working poor. Sub-Saharan Africa, on the other hand, is still lagging behind, with more than 50% of workers living on a dollar a day.

❖ **Target 3: Halve the proportion of people suffering from hunger (malnutrition, undernourishment) between 1990 and 2015.**

Rising commodity prices, but also food regimes, use in food regimes, urbanisation, the use of plots for production or the problem of subsidies to developed agriculture, make this target difficult to achieve.

South Asia and sub-Saharan Africa are the areas most in need, with under-nutrition caused by under-development.

1. **Ensuring primary education for all**

The goal is to ensure that, by 2015, everyone in the world will be able to complete a full course of primary schooling.

In 2006, 570 million children were in school, leaving 73 million children of school age out of school. 88% of children in developing countries are in school. This suggests that the target is achievable by 2015. In sub-Saharan Africa, the school enrolment rate was 12.5% in 2006, and in South Asia 9%. Experience shows that enrolment drops significantly when school fees are increased (as is the case in many African countries).

2. **Ensuring a sustainable human environment.**

This approach to environmental sustainability is based on 3 targets.

The first target is to integrate the principles of sustainable development into national policies and programmes and reverse the current trend towards the loss of natural resources;

The second target is to reduce biodiversity loss and achieve a significant reduction in the rate of biodiversity loss by 2010.

Although this target has not yet been reached, biodiversity will remain a global priority for at least the next 10 years, as demonstrated by the UN's declaration of 2011-2020 as the Decade of Biodiversity.

With a renewed strategy decided at the UN conference in Niagoya in 2010, and which will be specified at the Hyderabad conference on diversity in 2012.

The third target is to achieve a significant improvement in the lives of at least 100 million slum dwellers by 2020.

3. Building a global partnership for development.

- Official development assistance continues to fall, from a record of $107.1 billion in 2005 to $103.7 billion in 2007, yet each year an additional $18 million would have to be granted by the developed countries to achieve the doubling of aid decided by the West (G8) in 2015.
- Responding to the specific needs of the least developed countries, landlocked countries or areas and the smallest developing islands.
- Rapidly develop a more open commercial and financial system for electrification by promoting bilateral and multilateral cooperation worldwide.
- Sharing the benefits of the development of NICTs with developing countries

The number of subscribers to a fixed or mobile telephone has literally soared, from 530 million in 1990 to more than 4 billion at the end of 2006, including 2.7 billion for mobile telephones. This is a unique opportunity to bridge the technological gaps between poor and rich countries, with the mobile telephone often cited as one of the main tools for global development. Internet access will also help to meet several of the Millennium Development Goals.

STATE OF THE QUESTION

Our theme has already been tackled by several pioneers.

These are :

1. REC/INDE, rurale distribution and engry management en developing countries, March 2020 : The author concludes his scientific reflection by identifying the main causes of underdevelopment in rural areas, the different sources used to supply rural and urban-rural areas, determining the means of transmitting this energy to the consumer, and evaluating the billing method for rural consumption.
2. KHUZAMA and VIKOIS, the author of which has thought about the mechanisms that need to be put in place to ensure that the project is accepted in rural areas, where the population remains generally poor.
3. COASE, R le rôle de l'Etat (The role of the State), which deals with foundations and reforms, was concerned with determining the essential role of the State in supporting investors, who constitute a system with components such as territorial entities, communes and towns, political and administrative institutions, bodies or agencies, associations and public companies or enterprises that are generally non-active.

From all of the above, we will try to present certain managerial requirements (a dose of management) by presenting a model called the HNB, which claims to be a high-ranking manager with points such as effectiveness in management, efficiency and performance, profitability, performance and rationality.

3. ISSUES

It is vital that we work within the framework of the SDG objectives for the urban-rural population.

The constraints linked to the infrastructure are enormous: it has to ensure production, not forgetting transport and finally distribution to the population.

How can renewable energies be managed and used to develop rural or urban-rural areas?

- A very high investment cost for a dense, low-income population, taking socio-economic factors into account, would pose a complex problem in determining the real cost of the investment. How should the production of electrical energy be organised, taking account of socio-economic factors? What contribution can management make to the creation of a micro-hydropower plant?

In addition, very high energy costs will have an impact on the cost of agricultural produce exported to urban centres.

4. HYPOTHESIS OF THE WORK

The main causes of the lack of access to electrical energy would be a major preoccupation of local community residents, inefficient use of available resources, insufficient resources allocated to services. Inefficient distribution of resources between rural areas and wealthy populations, and high household expenditure make development difficult.

- The creation of a micro-hydropower station is an improvement and innovation aimed at alleviating the economic suffering of households at the height of the 21st century, and above all meeting the Millennium Development Goals for access to electrification and the development of rural areas by adding a dose of management.
- The hydropower plant can be seen as a vector for development and economic growth, so setting up a micro-hydropower plant requires participative management by the stakeholders:
- The reaction of activities that will generate development.

- The team (employees)
- Public institutions
- The local authority
- Private partners.

5. RESEARCH METHODS USED

All scientific research requires reference to world views shared by a scientific community known as an "epistemological paradigm".

The term paradigm refers to a constellation of beliefs, values and techniques shared by a given community (Kuhn, 1962).

Methodologically, the interpretivist epistemological paradigm framework is our choice, given that research into 'knowledge being better' in a complex structure such as the creation of a micropower station conditions not only the admissible research practices, but also the ways of justifying the knowledge developed, which must be applied in this work.
This is justified by the idea of Sandberg (a Swedish manufacturer of basic necessities) who maintains that different subjects participating in a certain situation are capable of agreeing on the attribution of a certain meaning by applying managerial culture. Using methods such as a survey sheet, a questioning survey and observation of the inhabitants, clearly identifying the needs of this community, with a view to establishing a diagnosis of the rationality, effectiveness, efficiency and performance of this structure in the commune of MALUKU, precisely in the KINZONO district, city and province of Kinshasa.

- This research is based on experiences in urban and rural areas.
- By observing the environment, we were able to identify the real needs of the population of the commune of Maluku, which is living in a disastrous situation.
- Contact with local residents in the Kinzono area and rewarding exchanges guided our approach.

I PART

CONCEPTUAL STUDY

CHAPTER I. GENERAL

INTRODUCTION

Disseminating scientific information is an intellectual operation, and one that is extremely difficult for both the research community and society as a whole. Few scientific products today can limit their field of activity to a single sector of human life. They are increasingly multidisciplinary and/or interdisciplinary, so that they can demonstrate their relevance.

Quite logically, information relating to the science of management is targeted because of its omnipresence in both public and private organisations, whether profit-making or not, in the market or non-market sector.

The science of management is a young discipline of our time, the contemporary era with very distant practices (prehistoric period). To be precise, between 1910 and 1925, the American Frederick Winolow Taylor, the Frenchman Henry Fayol and the German Max Weber put forward a series of conceptual and methodological proposals whose influence is still clearly perceptible today, both in practice and in our theories. Even if their motivations and approaches were different (Taylor was an engineer seeking to optimise structures, Fayol a manager wishing to bear witness to his experience and Weber the precursor of the sociology of organisations), we can legitimately consider them to be the founding fathers of the organisation and management, given that the latter remains a cross-disciplinary science.

Of course, managerial recipes have boosted the economies of companies and states around the world in their quest for results in terms of rationality, effectiveness, efficiency, relevance and performance.

In the light of the foregoing, this scientific exchange is taking place under the theme of general management and sectoral management, which is why we have divided it into four main areas: General management, the different types of management and the role of the sector.

sector management, strategic management and operational management.

Taylor used the expression "Scientifie management", translated as the scientific organisation of work, and his ideas on the administration of work can be found in Schap management public in the USA in 1902 and principales of scientif management, Paris 1911.

I.1. GENERAL MANAGEMENT

General management is a purist form of management, the preserve of business and the private sector. This is so true that we cannot doubt that management as a science has seen the light of day in business and industry. The primary seat of scientific management is the company. It is for this reason that this first section sets out to describe the components of general management.

1.1. Definitional aspects[2]

The greatest intellectual exercise that any researcher can undertake is to define concepts, in order to remove certain ambiguities. Indeed, Oscar NSAMAN-O-LUTU and Gode ANSHWEL OKEL MUTUNGI have argued that trying to define management is like asking the blind to touch the elephant and describe it. For them, it is perfectly logical that each blind person will define the elephant according to the part he has touched.

These two researchers simply compared management to the elephant and the various authors to the blind. This is to say that management is multi-disciplinary and multi-dimensional. Each author has tended to focus on his or her preferred sector when illuminating the concept of "management".

After studying several definitions of management, we have selected nine aspects to produce a very coherent definition that management researchers can agree on. These concepts are the foundations of management.

Management is an art, a science and a philosophy based on rationality, efficiency, effectiveness, performance and relevance (REEEPP). If we borrow the language of graphics, we would say :

Gode ANSHEL OKEL MUTINGI: Seminaire sur le management approfondie 2020 - 2022, Edition UNIC.

Figure 1: Definitional aspects of management

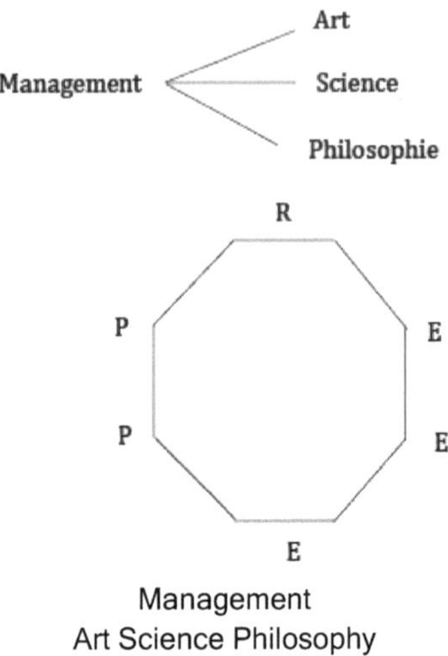

The 'art' of management means that it refers to what is innate, that it derives from innate intellectual capacities, from innate individual talents: it is an empirical form of management that is practised without accounting for it. Sylviane Friz traces the practice of management back to prehistoric times. Which is to say that management is not a "science". The practice of "management art" goes back a long way. Management science and management philosophy"[3] .

The object of management "science" or general management is the study of the company. This type of management originated in business. The most striking example is Fordism. Taylor's ideas enjoyed considerable international success, from Andre Citroc'n to Lenin. The most famous follower was the American Henry Ford. His aim was to dramatically increase the productivity of his car factories in order to produce middle-class workers. Scientific management" has two methods of analysing facts or managerial phenomena in vogue: systemics and strategy, for the simple reason that the company is considered to be a system and management is also considered to be a strategy. For a complete analysis of the managerial fact or phenomenon, general management resorts to live techniques (interviews, focus groups, observation, surveys) and non-living documentary techniques, such as social science photography.

Philosophical management is an ideology based on the promotion of moral and ethical values, a culture of results, success and excellence. Management

[3] O. NSAMAN - O - LUTU and G.ATSWEL - OKEL MUTUNGI, comprendre de management, culture, principe, outils et contingence, Kinshasa, ed. CAPM 2007.

philosophy goes hand in hand with management science and management art. It therefore focuses on promoting a managerial culture.
The six concepts or foundations of management (REEEPP) can be explained as follows: rationality, which is based on reason and is objective. For eating, it enables us to distinguish the useful from the pleasant. We believe that it enables us to transform the problem tree into a set of priorities (priority tree), a sign of intelligence and civilisation.
Effectiveness is the achievement of fixed objectives, efficiency is also the achievement of objectives but in record time and at lower cost. It means minimising costs and optimising time as an irrecoverable commodity. Effectiveness is the achievement of fixed objectives in their entirety or as a whole. Performance is the best-better or the best-more. Finally, relevance is the usefulness, the importance of each action, of each activity to be carried out by the manager, the useful action being the one that brings the results that lead to realism.
For a manager, the practice of management is based on the "OSR" trilogy, which stands for objectives-strategy-results, presented as follows:

Figure I.2: OSR trilogy

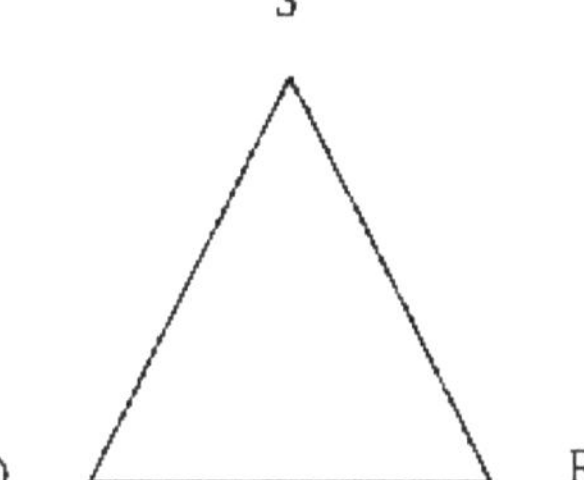

Commentary:
The first step in any managerial activity is to set objectives.
Next, strategies must be put in place to achieve results.
In management, what counts are results that express the achievement of objectives.

1.2.Managerial culture

After unveiling the nine definitional aspects of management, with three components that form the basis of its practice, we explore managerial culture, which is considered to be like the blood in the human body. It is through this that we have come to recognise managers[4] .
An intellectual is recognised by the fact that he reads, and managers also have managerial signs that are disseminated by the managerial culture.

4 S. FRITZ, Moi et le management, etre acteur de son propre developpement, Brussels Ed de Broeck, 1998. A DAYAN et al, ap cut p48.

In fact, managerial culture is made up of the indicators below that drive managerial behaviour.

Figure 3: Management culture indicator

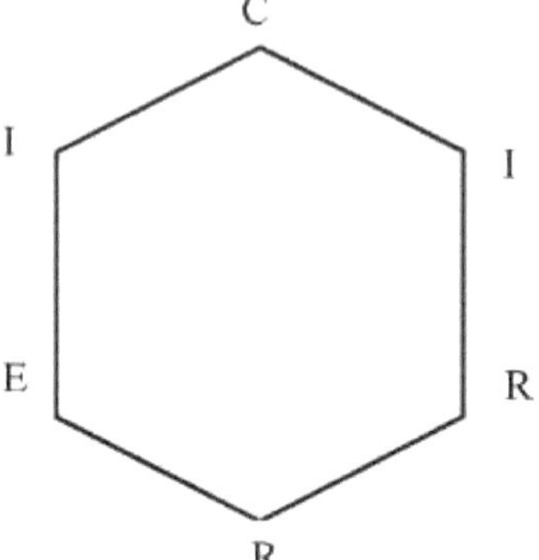

Comment: the management culture is based on creativity (C), innovation (I), entrepreneurship (E), calculated risk-taking (R), results (R) and entrepreneurial initiative.

Oscar NSAMAN - O - LUTU Oscar and Gode Atswel - Okel MUTUNGI have distinguished two types of managerial culture: masculine and feminine. The first, of Anglo-Saxon origin, is based on adaptation to working conditions, aggressiveness to achieve results, an excessive appetite for risk and the ability to act quickly to achieve objectives, and hard work combined with conflict.

On the other hand, the second, of Latin-Roman origin, focuses on the search for good working conditions, social peace and the avoidance of conflict.

It should be noted that, in practice, the masculine or feminine behaviour often adopted by managers depends on the context or situation.

1.3.Universal principles and tools for action

To be considered as such, apart from the object of study, the analysis methods and techniques must have universal management principles, which we symbolise scientifically as 'PROCOCOCODI', meaning forecasting/planning; organising; coordinating, commanding, controlling, directing.

The first five principles (PROCOCOCO) are called infinitives by Henry Fayol, who wrote them, while the sixth principle (DI) is the work of Peter Aniker, a lawyer and manager and the founding father of management by objective (DPO).

Forecasting or planning means looking into the future and drawing up a programme of action. Organising means setting up the company's two organisms, material and social, and establishing structures of authority; commanding means making things work, making people obey. Controlling means ensuring that everything is done in accordance with the established rules and the orders given. To lead is to give orders. To lead is to take decisions and make them operational.

Notwithstanding the universal principles mentioned above, if the manager does not use the tools below as a means of action, it will be difficult to achieve the objectives set.

These tools are marketing, IT, NICT, accounting, statistics, certain related sciences and above all the major management theories such as classical, neo-classical and human relations theories, not forgetting conflictualism and the decision-making process.

1.4.Sectors of activity

General management, or management as a science, was born in a company and consequently covets the "effects" of the company. These corporate effects concern human, financial and technical resources, which are managed by highlighting the resource of time.

In other words, areas such as HR management, financial management (financial and accounting management) and logistics management are all components that contribute to the performance of companies' sales.

2. SECTOR MANAGEMENT

It is a noble duty to point out that general management or scientific management, which was only operational in 1 company, has now seen the light of day in an organisation. The effort made is simply tropicalisation or domestication.

In this way, the manager is no longer just the company's Joker, the provider of the company's results or solutions, the doctor of the company but of any other organisation.

The transfer of management from the company to the organisation is explored here.

To be precise, the object of management study has broadened. At present, management specialists assert that according to the minimalist view, the object of management study is the company, whereas according to the maximalist view, the object of management study is the organisation.

The methods and techniques of analysis therefore depend on the field or sector sought or exposed by the management.

2.1.Management typology

The types of management we describe here clearly express the multi-dimensionality of management. This simply means applying the components of general management to the different organisational sectors of human life in society. It is a question of adaptability or pure and simple domestication.

To be more precise, sector management can be understood using the diagram below.

Figure 4: Types of management

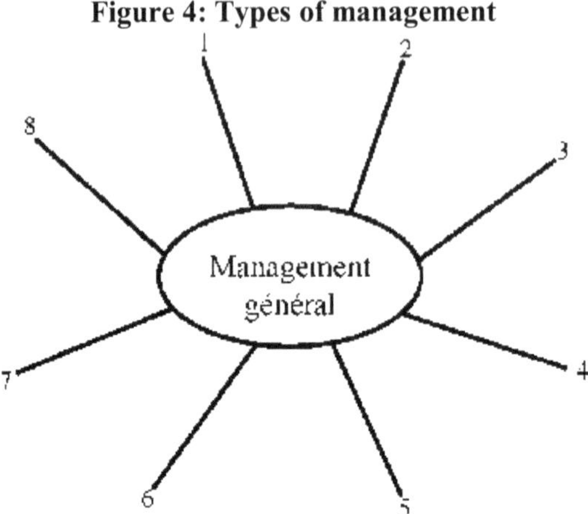

In the circle at the centre is general management, its components in fact. These universal components are transposed to X, Y and Z sectors, where they are domesticated/contextualised to produce the same effects as the company. In other words, each sector needs the components of general management to achieve its objectives in terms of rationality, effectiveness, efficiency, performance and relevance[5] .

To enlighten our readers, we have briefly outlined a sample of sector management based on the reference criteria mentioned below.

From this point of view, a distinction can be made between :

- Public management
- Territorial public management
- Political management
- Private management

Depending on the sector, we distinguish between :

- Economic management
- Public health management
- Development management
- Project management
- Church management
- Network and telecommunications development
- Meeting management
- Business unit management
- Service management

[5] MA. MORSAIN, Dictionnaire du management strategique, ED Belin sup, Paris, 2000, p 5 - 8

Organisational strategies include :

- Systemic management
- Strategic management
- Knowledge management
- Operational or tactical management
- Participative management
- Uncertainty management
- Conflict management
- Skills management
- Management by objectives

Without ignoring other types of management in the context of our strategic work.

3. STRATEGIC AND OPERATIONAL MANAGEMENT

According to OREAL, the difference between these two types of management stems essentially from the fact that strategic management is aimed at the long-term running of the company, whereas management that is rational...?

It is based on a triple logic (strategy formulation, execution and implementation of each strategic axis), whereas operational management focuses on the short term. It is concerned with solving the day-to-day problems of the company or organisation. Nevertheless, both less the notion of forecasting is made up of key functions constituting their common point in the planning system[6] .

Thus, in strategic management, forecasting extends over a very long period of five years or more. In operational management, the notion of forecasting, like all the other functions, supports the short term.

However, this differentiation does not mean that one form of management is used exclusively alongside another. They all complement each other and are used together, because the latter provides the technical support for the former insofar as it translates the exercise of strategic management into operational terms.

Moreover, we cannot think about the future without reference to the present. However, strategic management remains a variant of management that supplants operational management. It places the emphasis on strategy, on involving all members of the organisation in the planning and decision-making process. It is concerned with anticipating events and adapting the company to the context of an environment that is changing under the pressure of competition, technology and the demand for services.

---

[6] S. OREAL, Op.cit.p.55

CR. KICHMAN and M.A. SILVA, Testez-vous - memes votre entreprise, Paris, ed. Nouveau Horizon 1998. P.28.

I.3.1. Strategic management

Generally speaking, strategic management is in the hands of a company's senior executives, whereas operational management is the phase of strategy execution reserved for middle managers, whether they be marketing managers, sales managers, etc.

1. Definition

For I. ANSOFF, who coined the term in 1973, and certain theorists, strategic management is a paradigm, i.e. a specific scientific approach to the strategic object of a company.

Others, on the other hand, believe that it is simply a redundant name, since it would be obvious that management is strategic by nature.

According to G. KOEING, strategic management is an approach whose function is to ensure the competence, security and legitimacy of an organisation or system of collective action.

H. MITZBERG, for his part, defines strategic management as a method that strives for and is concerned with the implementation of intentions.

Mr MARCHESNAY believes that management is a discipline and process comprising five stages of industrial development, namely :

- Diagnosis;
- Strategic decision-making;
- Strategic formulation ;
- Strategic implementation
- Control and monitoring.

A. MARTINET defines strategic management as a discipline, a paradigm, a theory and a process comprising five stages in the development of an organisation, namely :

- Strategic diagnosis ;
- Strategic formulation ;
- Strategic action plan ;
- Strategic implementation ;
- Strategic steering or evolution of public action.

3.2. FOUNDATION OF STRATEGIC MANAGEMENT

Strategic management, as a method of steering a company, is essentially made up of several aspects.

3.2.1 Philosophical aspects

Strategic management is a way of thinking and acting that frees us from existing structures in order to avoid a conformist and straitjacket vision of the players involved, and to discern the most likely paths of development.

3.2.2 Operational aspect (Process)

Strategic management is based on a process known as strategic planning. Strategic planning is a continuous process by which the company determines and evaluates a priori the actions and decisions required to achieve a given objective[7] .

3.2.3 Ethical aspects

In the ethical aspects, these different elements will be developed in turn, in particular, corporate culture, leadership, management by objectives (DPO), motivation and organisational change.

I.4. OPERATIONAL MANAGEMENT

1. Definitions

Operational management is in line with strategic management, whose vision is for the medium and long term. The aim is to implement actions that are part of the day-to-day management of the company and consistent with its overall policy. Honestly, operational decisions have a rapid impact on the company's business, but they are reversible. For example, a change in the schedule, a repair to the operating system, etc.

These actions are limited and concern only some of the employees: they have no impact on the general running of the company. While senior management is responsible for strategic management, operational management is carried out on a day-to-day basis by the managers or middle managers of the various departments.

Operational management concerns the day-to-day running of a company. It involves all the techniques centred on the organisation of day-to-day activities, with objectives set in the short and medium term.

Operational management involves decisions taken by middle management (department heads, foremen, etc.) concerning the day-to-day running of the company.

These decisions concern the short or medium term and are aimed at optimising resources to achieve fixed objectives.

Examples of operational decisions.

- Setting up promotional campaigns
- Employee recruitment ;
- Pricing.

Operational management covers all decisions taken by managers for the day-to-day running of a company, department or shop. These are short or even

---

[7] G. KOEING, op. cit.
H. MINTZBERG, op. cit. p.122
A. MARCHESNAY, op.cit

medium-term decisions. Operational management implements the strategic policy defined by strategic management, which has a long-term vision.

Depending on the company's strategic objectives, it is necessary to :

- Setting short- and medium-term objectives;
- Identify the skills needed to achieve objectives and those already present within teams;
- Defining projects that fall within the scope of operational management.

2. Implementing operational management

Once the project has been defined, the operational manager is responsible for :

- Communicate the project to the whole team;
- Supporting the implementation of the plan by means of intermediate milestones, actions designed to motivate the team and by resolving any difficulties;
- Putting in place all the tools and resources to enable teams to work well.

This may involve specific equipment or collaborative tools for sharing information.

3. Operational management resources

To reach their destination, operational managers must therefore have access to all the information they need to fully understand the strategy. This will enable them to adapt the management tools as best they can, in line with the broad lines and decisions taken.

Operational managers have a team at their disposal. To manage his team, he needs to be able to keep them motivated by explaining the strategy and any deviations he observes. He will be able to rely on his team and manage them to find the solutions needed to achieve the objectives set.

The operational manager must have all the autonomy needed to make changes. Otherwise, if the strategy is too rigid, it may be difficult to adapt, and ultimately difficult to make the decisions needed to ensure operational effectiveness.

This is often the criticism levelled by operational managers at human resources management. They often lack autonomy and cannot make decisions to adapt to operational objectives.

Which is a tragic mistake, because strategic management has to leave him in charge. The strategic operational manager must leave him in charge. The operational manager must therefore have the confidence of his or her line manager. Otherwise, as the operational manager, he or she will not be able to create all the value for which he or she is paid.

This person is involved in local management. They are at the heart of interactions between teams, suppliers and other departments. Because it's only by ensuring overall coherence that he can adapt the strategy to the realities on the ground.

The operational manager is therefore not simply an executor. On the contrary, the job of operational manager consists of guaranteeing the operational framework. The operational manager also ensures that the strategy is properly implemented in a context where adaptation is essential.

4. Creating value in operational management

Operational management, because it adapts the strategy to the situation on the ground.

Creating significant value :

They ensure that the company runs smoothly on a day-to-day basis, offering customers the best possible products and services. They ensure that customer satisfaction is continuously improved by adapting to their needs and expectations;

The operational manager, because he can be in contact with suppliers to ensure that products or services are available. In this way, he or she guarantees fluidity of supply and ensures that stocks are consistent;

The operational manager is therefore an essential link in guaranteeing operational feasibility.

Its role is also to pass on information such as problems encountered or unavoidable needs.

It therefore communicates with central services and top management. This enables them to get things moving so that products, services, communications and budgets are aligned with their problems and needs;

The operational manager ensures that teams are fully motivated and involved. This is because of the various problems, the many changes and the loss of understanding of what they do. By ensuring that they give meaning and autonomy despite the mismatch between strategy and the field, they help to develop skills and maintain a high level of commitment from their teams. As a result, they can fully create value.

CHAPTER II. THE CONCEPT OF CREATIVITY

11.1. THEORETICAL CONCEPT[10]

Since the dawn of time, the notion of creation has been a key factor in development and the creation of wealth to combat poverty throughout the world, with a particular focus on business creation.

The term "company" is commonly used today, especially in the business world. It is also used to describe many different professions, including farming, industry, the private sector and the public sector. Where does this brand name for businesses come from?

First of all, by definition, a company is an autonomous economic unit that produces goods and services through sales and distributes income. Moreover, our contemporary world is dominated by activities of a commercial nature.

On another level, in a good number of developing countries, unemployment is a definitive decent job or ensuring decent employment for all jobseekers. Governments support private-sector initiatives to create jobs and self-employment, especially in the important field of management.

In practice, every citizen is called upon to become their own employer. For the students in particular, who will be completing their 3^{e} cycle in the next few days, the solution for the future lies in creating their own jobs.

These are just some of the factors that explain why the company has become a real subject of interest and attraction.

Of course, it is not possible to cover the entire subject of simple communication.

11.2. THE CONCEPT OF BUSINESS, THE ENTREPRENEUR, ENTREPRENEURSHIP AND THE ENTREPRENEURIAL SPIRIT

The term 'company' is used very frequently in the business world, without it being possible to agree on a single definition. Nowadays, however, it is extremely important to agree on the content to be given to the concept of the company at national, regional and international level, and it is difficult to resolve certain issues such as taxation or the general way of promoting the development of the company.

France's National Institute of Statistics and Economics (INSEE) defines a company as a legally autonomous economic entity organised to produce goods or services for the market.

In practice, companies come in all shapes and sizes, with different legal forms and belonging to different professional sectors.

According to FRANCOIS PERROUX, an enterprise is a form of production in

---

[10] FAYOL Alain, entrepreneurship, learning to undertake 2nd edition, Dunold, Paris, 2007
APCE (Agence pour la creation d'entreprendre) creer ou entreprendre une entreprise, 13th edition, Paris 2000.

which, within a single asset, the prices of the various factors of production contributed by agents other than the owner of the enterprise are combined in order to sell goods or services on the market and to obtain a monetary income resulting from the difference between two price series.

According to TO - MORROWIS BUSINESS, in New - York, a business is an organisation of individuals of varying abilities using capital and talent to produce some good.

According to STEPHANIE BALLANDE and Anne Marie BOUVIER (2019), the company is a multidimensional entity: economic and social, which can be sold for more than it costs.

As for LANZEL in the chart of accounts, it is a hierarchy that uses intellectual, physical and financial resources to extract and transform information, in accordance with defined objectives, by a Personal and Collegial Management, involving, to varying degrees, profit and social use motivations.

CAMION, in its treatise on private enterprise, considers that an enterprise is 'a financially independent body whose purpose is to produce certain goods or services for the market'.

Furthermore, M le BRETON (Business Economist) prefers the term organisation instead of organism because he maintains that the enterprise is "any form of economic organisation, financially autonomous, which sets out to produce a good or service for the market".

To give a precise content to this notion, we use the definition given by the economic lexicon. Le vocabulaire et les mecanismes de l'economie' published by Cyril GOUANGOUNGOU in 2020.

A company is any natural or legal person who, with the aim of making a profit, optimally combines labour and capital to produce a product to satisfy a solvent demand expressed on a market.

Finally, it is also important not only to cite its authors and researchers, but also to identify certain characteristics of the company as an economic organisation that are essentially common to all companies.

11.2.1 Company characteristics

All companies can only be :

1. **An organisation**: i.e. a company is a stable, lasting organisation that has been given a certain structure, in order to carry out a series of operations.

There is no need to create an isolated act, as the business involves a series of operations that are usually repeated.

An economic player :

- It is a production unit that transforms inputs into products and services,
- It contributes to the formation of gross domestic product (GDP) by

generating value added (VA);

- It is also a unit of expenditure that consumes and invests to ensure production;
- It is a unit for distributing and sharing wealth;
- Because of its commercial nature, it is subject to constraints of effectiveness (meeting fixed objectives) and efficiency (meeting objectives while optimising resources).

2. Human reality

- The company is also defined as a community, a human group, employees who contribute to the achievement of common strategic objectives;
- Individuals must work together to achieve their objectives;
- The interests of the company and the interests of the individual must converge.

3. Social reality

- The company creates jobs, income and products, but also innovation and technological progress;
- The company acts on its environment, its activity having repercussions on those of other economic agents;
- Companies take certain actions spontaneously or under pressure from the environment.

4. Autonomous financing: This means that, in theory, a company that organises its initial financing as it sees fit does not constitute a company, since it is not autonomously financed, but merely an establishment. However, it should be noted that the company's financial autonomy may be limited in certain cases if it is part of a trust, cartel, agreement, etc.

5. Working for the market: In other words, it works to find and satisfy customers (satisfaction management). It is at this level that the nation of business risk appears, where the entrepreneur is the one who undertakes, whose economic action is subject to certain uncertainties because he is an economic agent who assumes technical and commercial risks.

6. Produces goods or services: i.e. it can produce consumer goods that will be destroyed by the first use or production goods that are durable and can be used in a new production cycle. Examples include banks, insurance companies, etc.

SCHUMPETER, an American economist, and GASTON PETER presented two major characteristics of a modern company:

SCHUMP PETER has highlighted the dynamic nature of the company, arguing that the essential function of the company is to relaunch and maintain economic expansion through innovation and new combinations. It should be noted that the company, from its conception to the manufacture and marketing of a new

product, must undergo change every time.
The introduction of a new method of production, organisation or distribution (assembly line or series production, prefabrication in the building industry, etc.).
The opening up of a new economic opportunity in the case of the sale of plants set up for trading. The conquest of a new raw material or energy source, for example the discovery and extraction of gas or oil.
On the other hand, Gaston PETER, in the forward-looking nature of the company, insists that its lasting success depends on the accuracy of its short- and long-term forecasts, i.e. in the broadest sense, on the accuracy of its anticipations.
In order to achieve this, the company has to identify future needs and prepare to meet them, which requires a clear change in the company's structure and methods. In fact, the increasing importance of research, scientific study of markets (commercial services), the development of forward-looking management methods and, finally, the development of a training, information and retraining programme for staff at all levels.
The modern company must therefore be progressive by pursuing very specific objectives.

11.2.2. Company roles

In order to set up a business as a coordinator of the factors of production, the entrepreneur must :

- Seeking to raise the necessary capital;
- Choosing a location, premises and equipment;
- Recruiting and training the workforce ;
- Organising production and sales.

In addition, the contractor must take into account the following actions in particular:

- Attraction
- Conversion
- Production
- Delivery

To do all this, he has to distinguish between the company's roles vis-à-vis its customers, its agents and its owners.

11.2.3The company's role in relation - to customers

The company is not a centre of internal production and exchange with the outside world. That's why a company normally looks to the outside world, to the customers it needs, or else it will disappear[11] .
The true purpose of the modern company is not to multiply consumer goods ad

[11] BIZARGUET. A, le secteur public et la privatisation, Paris, PUF, 1980, p52

infinitum and satisfy needs that are often artificially created by external advertising or marketing, but to ensure the harmonious development of an economy geared towards satisfying, as a priority, needs that are recognised as being the most useful from a human and social point of view.

11.2.4The company's role with regard - to its employees

The company provides a useful activity called the company's social role (CSR) for all those involved in its operation.

At the same time, it must provide them with a social situation suited to their abilities. In this case :

- The necessary technical resources and the general organisation within the company to enable staff to achieve the best industrial and commercial performance;
- A working environment, an atmosphere where staff enjoy working, living and thriving;
- It must therefore be considered from both a material and a psychological point of view;
- A decent salary with purchasing power;
- A degree of job security. It is not a question of systematically keeping a given job, but of transferring the working population, with the necessary economic progress, from one activity to another within the company.

11.2.5Role of the company vis-à-vis the owners of the business

By owner, we mean the operator in the case of a sole proprietorship, the shareholder in the case of a joint-stock company, and the state in the case of an overall profit. In other words, the company must make a profit, and this is just as necessary in a collectivist system as in a capitalist one.

To ensure that the business does not fall to the community. So a loss-making or bankrupt company brings losses and suffering to its unpaid staff, and when it's an important company that can't be allowed to disappear, it's up to the community as a whole to make it possible for the company to finance itself, called accumulation, under the collectivist system;

Failing that, self-financing and business expansion will have to be achieved through private or public savings, by calling on public or private banks.

11.2.6Role vis-a-vis the general economy of the country оù it is located

When a company's importance and the volume of its operations are such that the life of the country depends on it, the managers of that company cannot ignore their responsibility to the country as a whole, because their decisions can have a considerable impact[12] .

[12] AURELIEN et AL, theorie economique, introduction aux theories de l'etat, Paris, ed. PUF, 2012, p156.

When all is said and done, a company has a dual role to play: **on the** one hand, **it makes** an **economic contribution**, i.e. **it** makes a product, a commodity or a service available to the public; on the other hand, it **makes a social contribution**, insofar as an economy must be established for the benefit of mankind. Every company is a working community in which people at all levels work together, aware of their solidarity and concerned for the interests of the community.

So a company must be both a means of individual enrichment for those who contribute their labour or capital, and a means of collective enrichment for those who created it and the beneficiaries of its products and services.

II.2.6.1. Business Sectors

Let's remember that the **essential** purpose of a company is not to be a simple economic unit whose sole aim is to produce and sell goods and services in order to earn a monetary income, the profit[13] .

It is, of course, the place where people spend practically half their working lives, the other half being devoted to family life and leisure activities. Every company must therefore be at the service of man, its creator and its driving force. It must therefore create wealth in terms of money, culture and morals.

These sectors enable companies to develop wealth classified as :

1. Material wealth

This wealth, once created, will have to benefit :

- Shareholders who have invested, because the profits generated will have to be distributed to shareholders in proportion to their shareholding and also at the risk of legitimately recognising this production unit in the eyes of capital investors.
- Employees who have worked because the quality of the product and the volume of the profit are directly proportional to the quality of the human resources and the work provided. It is only fair that the staff who provided the labour should share in the profit generated. This wealth can also benefit employees in the form of remuneration by improving working and living conditions or by ensuring greater security of existence;
- Customers who have bought because the increase in wealth in the company is a function of its clientele, in particular the higher quality of goods and services sold in relation to the many needs expressed by consumers;
- Suppliers who have sold raw materials for production or external services necessary for the management of the business or the well-being of the staff are, of course, the means by which suppliers become richer,

---

BERNOUX, P. La sociologie des organisations, Paris du seuil, 2006, p 32 - 38.

[13] BRUNS and STACKER, the management of innovation, Oxford, ed. OUP. Oxford. SD

- National community that collects taxes and other social contributions.
- Through the levy, the State in whatever form, the company participates in a national solidarity effort with a view to the construction of infrastructure needed by all, the implementation of useful and necessary means for the enrichment and safeguarding of the community and ultimately to help the most disadvantaged and the security of harmonious existence of everyone in the community.

2. Cultural wealth

Here, the company, being the place where, as we have said, men and women who live there for eight hours a day according to the current standard of work come together, deserves to be described as the nest of functions for which these men and women give their efforts and their vitality.

It should therefore enable the full and harmonious development of the whole intellectual personality, helping each individual in all circumstances to form sound judgements and develop a critical mind.

As a cultural mosaic, the company must take a positive approach to any technical innovation, which has certain repercussions on the entire culture of the people who live there, live alongside it and depend on it, because sometimes those who use its techniques or products benefit from the cultural wealth it generates.

3. Social Wealth

Is it during a whole existence of the company, men and women meet, work, compete in the company, are confronted every day with questions of justice, equality, honesty. All this in the daily darkness and tension of active work. They have to relax: and make an effort to understand each other.

If they know how to resolve these latent conflicts themselves or with the help of those in charge. In the final analysis, the company is able to create a high level of moral wealth, because it is active and put to the test every day.

II.2.6.2. Company typology

This typology is established on the basis of certain elements, including the object of the activity, the objectives pursued, the shapes and dimensions.

Depending on the purpose of the business, a distinction is made between agricultural, industrial, commercial and service businesses, such as transport, banking, insurance, hotels, tourism, catering and entertainment or leisure businesses;

Depending on the objectives pursued, a distinction needs to be made between companies that are "riveted", concerned above all with making a profit, and public companies where the profit motive is not systematically proven and where what matters above all is the management of a public service in the general interest;

Depending on the legal forms taken by the companies, it would be better to speak here of a variety and diversity that deals with companies: in corporate companies (partnerships, general partnerships, limited partnerships and joint stock companies), in joint stock companies (limited partnerships with shares, cooperative companies, nationalised companies, mixed economy companies);
Depending on the size of the business, we distinguish between small businesses, which we call small craftsmen or tradesmen, where the entrepreneur, while running his business, takes an active part in the execution of the work; in this case we speak of Partisan and small tradesman; medium-sized companies, where control of the work is often carried out by the entrepreneur himself, with a staff of at least 50 people; and finally large companies, where control is carried out by a technical director, while the entrepreneur reserves commercial and financial management for himself, with a staff of at least 100 people, and where a board of directors reserves senior management and important decisions for itself.

II.3. BUSINESS FUNCTIONS

Different functions enable a company to achieve its objectives, including administrative, accounting, financial, commercial, marketing, social and technical functions.

The administrative function is used to harmonise the operation of the various departments by coordinating their activities and controlling them. It is the brain of the company, because it enables the other six functions to be directed and managed, and is therefore the responsibility of the Chief Executive.

Whereas the accounting and financial function, on the one hand, makes it possible to forecast and search for or manage financial resources, its role is to control forecasts, not to be a body for recording past events, as was once the case with a static body; In addition, its role is to record facts in order to establish the company's situation and results at the time of the balance sheet. Modern accounting has become a dynamic body which must be able to provide a great deal of information in the form of statistics and must be able to transmit this information rapidly using an appropriate accounting technique, or else, when it must be able to lead to the establishment of programme budgets, sales budgets or purchasing budgets,

In addition, the finance department is responsible for solving the company's financing problems and must therefore enable it to raise the funds it needs, whether in the short or long term, and must also ensure that this capital is used in the most profitable way.

However, the commercial function covers all the activities of a company relating to commercial exchanges. It is an intermediary function that handles relations with customers on the one hand and suppliers on the other. This

function creates the flow of products on which profit depends. In the final analysis, it is essential, because it is on this that prosperity depends.
In this way, the IT function brings together a mass of useful information 4 which is growing day by day and which is constantly being obtained from different and diversified data, giving rise to a new science called computer science, based on the use of computers.
The function of Marketing is that which enables any company whose purpose is to produce goods and services for the market and to adapt to it by the means available which influence the company favourably. Communication, sales, the market, the product or service and packaging are important Marketing concepts. The study of the needs and psychology of the consumer and the means of influencing them constitute the Marketing in question.
Finally, the social function is one of the functions born of the company in the last few years or FAYOL H. had expressly indicated it. This is why we sometimes see it attached to another function, such as the safety and hygiene function, which is an extension of the personal safety function found either in the administrative function as an extension of personnel administration. This function determines the types of products to be purchased and the work to be carried out, which in turn determines the raw materials and consumables to be supplied.
From these functions, we can classify companies in several ways according to their legal nature, their size and their field of activity.
We need to distinguish between :

1. **From the point of view of form**

We distinguish between sole traders and partnerships.

Sole proprietorships are those where the source of funds is a person, an individual, a capitalist and the sole master of the business, owning the absolute initiative and the sole holder of the capital, whereas companies are owned by people who are known as partners, who contribute capital, are legally responsible and receive a variable remuneration. It should be noted that a company is a contract by which two or more individuals agree to pool their goods or services in order to share in the results.
The business tradition encompasses several types of company
which are :
Partnerships include the following types of commercial company: sociétés en nom collectif (SNC) and sociétés en commandite simple (SCS),

Capital companies.

There is one particular characteristic: "the personality of the partners or the fact that they have little to do with the company, but it is their contributions that play a role in the limited liability companies (SARLs) and cooperative societies.

2. From the point of view of their sector of activity

We distinguish between companies in the primary sector, which exploit a natural element: mining, quarrying, fishing, agriculture, forestry, oil extraction, delivering consumer goods without relying on the transformation of raw materials; companies in the secondary sector, which are industrial companies that transform raw materials into finished products to serve consumers; and finally companies in the tertiary sector, which make their services or finished products available to customers as part of the administration of commercial affairs.

3. In terms of size

Their businesses are differentiated according to the volume of capital invested, the size of the business, the number of employees and its complexity, in particular: small and medium-sized enterprises (SMEs), which include modern businesses that employ at least fifty people in purely family activities, village industry activities and associations that operate micro-businesses in unstructured sectors of the economy, and large businesses (multinationals), which carry out and control production operations in several countries outside their country of origin.

4. From the point of view of their regime.

There are three categories of company: public companies, private companies and mixed or parastatal companies.

It should be noted that private companies are owned by private individuals who set them up, manage them and own their assets; mixed companies are legal entities and own their assets, the capital of which is made up of contributions from private individuals; public companies, on the other hand, are the sole property of the State.

With regard to this point of view, our particularity is to be interested only in the public companies of the DRC whose five important sectors are:

- The energy sector (Societe Nationale d'Electricite, "SNEL", Regie de Distribution d'Eau "REGIDESO"). These companies have a monopoly on the production and distribution of electricity and water,
- The mining sector (Generale des Carrieres et des Mines " GECAMINES la Miniere de Bakwanga " MIBA " et les autres qui sont en situation de monopole aussi),
- The Post and Telecommunication sector (OCPT),
- The insurance and public health services sector,
- The transport sector (the former Office national des transports "ONATRA", the societe commerciale des ports et des transports "SCPT", the Societe

Nationale des Chemins de Fer au Congo "SNCC", the Etablissement de transport au Congo "TRANSCO", etc.).

11.4. PUBLIC COMPANIES IN THE DRC

There are many thinkers, researchers and authors who have given meaning to public companies.

Definitions are not of capital importance in what we are dealing with, however, to avoid doctrinalization and by the absence of an official definition which incite authors to controversy and to take varied and divergent positions, the notion of public companies those called private, will allow us to distinguish and to see clearly on the qualifier "Public" added to the word company.

This gives rise to the comparison between them that we would not know without placing them in the three elements that are :

- The aim;
- Ownership,
- Company management.

Without, however, going into the details of the matter, the aim pursued by public companies is still the general interest, whereas the aim of private companies is profit or private interest, because the State does not make a profit when it sets up public companies, as we can see from the other two elements.

In fact, a public enterprise is an economic and legal unit in the same way as a private enterprise in that it is a coordinated operation with human and material resources to ensure the production and distribution of economic goods and services and has a legal personality, i.e. it is subject to the law, On the other hand, public companies are not subject to the same economic principles as private companies, because they are not on an equal footing with private companies in terms of credit, taxation, the treatment of their staff, their assets and their liability, as well as the objectives pursued.

Private companies are primarily concerned with "financial profit", whereas in the case of public companies, the aim is to serve the public interest.

As far as ownership is concerned, the State owns public companies because it decides, organises, recruits and appoints managers as "Public Representatives", whereas private companies are owned by third parties who have invested their capital in the company for which they are the main author.

Table 1. Typology of legal types of company

Types& characteristics	**Sole proprietorship**	**Company people**	**Societea limited liability**	**Company capital**
Contributors	Sole proprietorship	Associates	Associates	Shareholders
Elements pooled		Personalities and skills of the partners	Personalities and skills of the partners	Capital

Responsibility for 1. debts	Unlimited on the personal property of the...	Unlimited and for each partner	Unlimited to contributions	Unlimited to contributions
Name of shares held by contributors	No	Company shares	Company shares	Action
Management	Contractor	Manager(s)	Manager(s)	Council
				management

Source: MUKENDI MUNTU Pierre - Espoir dans le management de projet dans les entreprises et les organisations, Approche generale et perspective.

In the case of a private company, it is the entrepreneur who decides whether to run his company alone or to entrust it to a manager who remains accountable to him, as shown in table 1.

Some other authors have defined companies and have contributed fagon source which we call the contribution according to them of the comprehension on the public or private companies in legal term and others. But before reviewing their thinking, we wanted to centralize our research by focusing the approach around the recent reform on public enterprises after the reform of 07 July 2008.

Section 1. Public Companies after the 07 July 2008 reform

The reform of public companies on 07 July 2008 is based primarily on the State's desire to revitalise public companies with a view to making them efficient and profitable. This reform led to the creation of the following laws[14] :

- Law no 08/007 of 07 July 2008 on general provisions relating to the transformation of public companies;
- Law 11⁰ 08/008 of 07 July 2008 laying down general provisions relating to the withdrawal of the State from portfolio companies;
- Law no.⁰ 08/009 of 07 July 2008 laying down general provisions applicable to public establishments;
- Law no 08/010 of 07 July 2008 laying down the rules relating to the organisation and management of the State portfolio concerning the measures for implementing the said laws enacted by decrees no° 09/11, no 09/12, no° 09/13, no° 13/14 and no° 09/15 of 24 April 2009 laying down the legal, economic and financial measures necessary for the implementation of the transformation of companies as well as their organisation and operation into commercial companies, public establishments and public services until the date of adoption or determination of their respective statutes are distributed and grouped according to the different lists in the tables below:

Table 2. List of companies transformed into commercial companies

Sectors	Denomination	Code

[14] Reform of 07 July concerning the transformation of public companies into commercial companies, public establishments or public services and public companies and public companies to be wound up.

Mining	General quarries and mines	GECAMINES
	Company developpemen industrial and mining Congo	SODIMICO
	Office des Mines d'Or d Kilomoto	OKIMO
Energetic	Water distribution company	REGIDESO
	Societe Nationale d'Electricite	SNEL
	Congolaise des Hydrocarbures	COHYDRO
Industrial	Societe Surdergique de Maluku	SOSIDER
	Societe Africaine d'Explosifs	AFRIDEX
Transport	Societe Nationale de Chemin d Fer du Congo	SNCC
	Office National des Transports	ONATRA
	Regie des Voies Aeriennes	RVA
	Regie des Voies Maritimes	RVM
	Congolese Airlines	LAKE
	Compagnie Maritime du Congo	CMDC
	Chemin de Fer des Uelles	CFU
Telecommunication	Office Congolais des Postes e Telecommunications	OCPT
Finance	Caisse d'Epargne du Congo	CADECO
	Societe Nationale d'Assurances	SONAS
Service	Hotel Karavia	KARAVIA

Source: Ministere de Portefeuille, Conseil Supërieur du Portefeuille, Rapport d'activite 2001.

Table 3. List of public companies transformed into public establishments

Sectors	**Denomination**	**Code**
Agriculture	Office National de Cafe	ONC
Transport	Regie des Voies Fluviales	RVF
	Fre Management Office Maritime	OGEFREM
	City - Train	CITY - TRAIN
T elecommunication	Press Promotion Agency	ACP
	Radio Television National Congolaise	RTNC
Financial	Fond de Promotion de 1 Industrie	REIT
	Caisse Nationale de Securit Social	CNSS
Construction	Office de Route	OR

	Office des Voies et Drainage	OVD
Service	National Tourist Office	ONT
	Office de Promotion des PME d Congo	OPEC
Trade	Kinshasa International Fair	FIKIN
	Office Congolais de Controle	OCC
Search	National Institute of Statistics	INS
	National Institute for Agricultural Studies and Research	INERA
Nature conservation	Institut Congolais pour l Nature conservation	ICCN
	Institute of Zoological and Botanical Gardens of Congo	IJZBC
		IMNC
Training	Institut National de Preparatio: Professionnelle	INPP

Source : meme source

Table 4. List of public companies transformed into public services

Sectors	**Denomination**	**Code**
Agriculture	Office National de Developpement de 1'Elevage (National Office for Livestock Development)	ONDE
Mining	Centre d'Expertise d'Evaluation e de Certificat des Substances Minerales Precieuses et Semi - Precieuses (Centre for Expertise, Evaluation and Certification of Precious and Semi-Precious Mineral Substances)	CEEC
Financial	Office de Gestion de la Dette Publique	
	Directorate General of Customs and Excise	DGDA
Service	Regie National d'Approvisionnement et de l'Imprimerie	RENAPI

Source : deja citee

Table 5. List of public companies to be wound up

Sectors	Denomination	Code
Agricultural	Sulu cocoa	CACAOCO
	Gosuma palm grove	PALMECO
	Cotonniere du Congo	COTONGO
	Lotokia Sugar Complex	CSC
	Caisse de Stabilisation Cotonniere	CSCO
Service	Office des Biens Mal Acquis	OBAMA

Source : meme source

It should be noted that this reform has been devoted more to the probiematique structures of companies as we have emphasized in our studies without taking care of the man who must implement these structures as the main animator of all the stakes, that is why, Some authors have drawn up strategies for managing private companies and for transforming them into the management of public companies, namely :

- . Ludovic BERTHIER et al. believe that public management draws its strengths and strategies from private management, so we have to understand that it is from private companies that the public approach to managing public companies draws.

In short, they distinguish between two types of company: public and private, where management strategies must not be copied but adapted and fine-tuned to achieve objectives.

- . Other researchers such as Andre CABANIS, Pierre MULLER, Laurent RICHER and others, whose list is non-exhaustive, help us to master the entrepreneurial environment and to distinguish a threefold typology of enterprises: public enterprises in which the State is the sole principal actor in its creation and management; private enterprises created and managed by actors other than the State, individually or collectively; and finally, parastatal enterprises or mixed enterprises, which are hybrid or combined forms of enterprise in their creation and management.

In this case, the businesses created must be supported by innovation and collective experimentation in a public-private partnership approach that enriches the interplay between the players involved, taking account of their ethical and collective values and remembering their different capacities to seize opportunities, their ulterior motives, their corporate or community culture and their rational anticipation and futurology.

In order to better understand these concepts of public companies, the reform we have just discussed has used terms such as public company for a large number of companies whose aim is to make their activities commercial by withdrawing their mission to pursue the general interest because they must make profits on behalf of the company whose maximisation of profits is the only behaviour that affects the initiators (the State), Public establishments or public services in the material sense are all companies whose activity is intended to satisfy needs in the general interest and which, as such, must be ensured or controlled by the supervisory administration because they are integrated into the administrations of the ministries concerned.

Hence, we can no longer begin to understand what we call the transport sector as an expression that simply indicates the operation by air, sea, river and road of companies that deal with the mobility of people and their goods in urban areas or between provinces or between regional or state administrative entities (cities of two States).

In our work, the term public transport sector refers to the sector concerned with transporting people and their goods by road.

This clarification allows us to start with the concepts that are part of the model concepts that we are proposing for the Manager to be called Manager and Leader.

Section 2. ENTREPRENEURSHIP

1. Definition

Another term much used in this context is entrepreneurship. Entrepreneurship is defined as the process by which a person or group of people put time and capital into the search for market opportunities, with a view to generating value and growing the business through innovation, whatever the resources available.

It is the process of launching a project, organising the necessary resources and insuring both the risks and the benefits[13] .[15]

2. The entrepreneur and the entrepreneurial spirit

❖ **Definition**

But what about the "entrepreneur", a term that is often used in everyday life and particularly in the business world?

The entrepreneur is the person who sets up or runs the business. He or she has innovative working skills.

Famous authors have given their take on the entrepreneur:

- For SCHUMPETER (1995), an entrepreneur is a person who is willing and able to transform an idea or invention into a successful innovation.
- According to Peter DRUKER (1970), an entrepreneur is someone who is prepared to put his or her career, time and capital on the line to realise an idea, sometimes even under risky conditions.

In the vocabulary used to describe a company director, terms such as courage,

[15] AU DE COSTER Michel, sociologie du travail et gestion du personnel, collection gestion et organisation des entreprises, Ed Labor Bruxelles, 1976.

invention, perseverance, initiative, risk-taking, etc. are often used.
We will end this series of definitions with the notion of entrepreneurship.
The Centre d'analyse des politiques economiques et sociales (CAPES) in Burkina Faso, in its December 2004 publication entitled 'Les fondements de l'entreprenariat en RDC', defines entrepreneurship as 'the creative ability of the individual, alone or within an organisation, to identify an opportunity and seize it to produce new value,...
In other words, it's the ability of a person or a group of people to embark on some kind of adventure to create something new, with all the risks that this can entail.

II.5. INTRODUCTION TO BUSINESS CREATION

The second edition of the University Days organised by the Anselme Titianma SANON Foundation for Culture, Peace and Development (FATISA) is focusing on the theme of Generation worthy of meeting today's challenges. Its aim is to help young people prepare for the future by developing their knowledge, skills and attitudes.
We've chosen to make our modest contribution to this with a communication on business creation. Why such a subject?
The term "company" is commonly used today, especially in the business world. It is also used to describe many different professions (agricultural enterprises, industrial enterprises, financial enterprises, etc.) and in different forms (private enterprises and public enterprises). Where does this marked interest in business come from?
First of all, since the company is by definition an autonomous economic unit producing goods and services for sale and distributing income, it is bound to attract the attention of every citizen. Moreover, our contemporary world is dominated by commercial activities.
On another level, we can see that in many countries, both developed and developing, unemployment is a worrying problem for governments. Faced with the impossibility of curbing this problem once and for all, or of ensuring decent employment for all jobseekers, governments are supporting private initiatives to create jobs and, above all, self-employment. In practice, every citizen is called upon to become his or her own employer. For young people in particular, who have just finished university or vocational schools, the solution for the future lies in creating their own jobs, which is why they are so keen to run their own businesses.
These are just some of the factors that explain why the company has become a real subject of interest and attraction.
Of course, it's not possible to cover the whole subject in a single

communication. You could devote an entire book and practical exercises to it. For all these reasons, we have entitled this paper "Introduction to Business Creation". It is structured around the following points:

- The concepts of business, the entrepreneur, entrepreneurship and the entrepreneurial spirit

;

- The preliminary work involved in setting up a business;
- Formalities for setting up a business in the DRC.

With this approach, which is all in all synthetic, we hope to meet the information needs of young and future entrepreneurs.

II.5.1. Notions of enterprise, the entrepreneur, entrepreneurship and the entrepreneurial spirit entrepreneurship

11.5.1.1. Definitions of business and entrepreneurship

France's National Institute of Statistics and Economics (INSEE) defines a company as "a legally autonomous economic unit organised to produce goods or services for the market".

In practice, companies come in all shapes and sizes, have different legal forms and belong to different professional sectors. Hence the difficulty of a single definition of a company.

A **company** is any natural or legal person who, with the aim of making a **profit,** combines labour and capital in the best possible way to produce a product designed to satisfy a solvent demand expressed on a market.

A company can take several forms. It may comprise one or more establishments which are technically autonomous production centres but are economically and legally integrated into the undertaking. It may be set up and owned by one or more natural or legal persons.

A distinction is made between :

- Private companies set up by private individuals or legal entities;
- Public companies are those whose owners include at least one person governed by public law, i.e. the State or one of its agencies.
- MEMBERSHIPS ;
- Semi-private or semi-public companies that are halfway between a private company and a public company.

There are two forms of private enterprise: sole proprietorship and company.

A sole trader is one whose company name, assets and income are the same as those of the household. The household is then the owner. Examples include shopkeepers, farmers, craftsmen, lawyers, bailiffs, etc.

In the case of a company, the capital is divided between several natural or legal persons. Companies are divided into partnerships and limited companies. Partnerships are those in which the named partners hold shares in the capital which cannot be sold or transferred without the unanimous agreement of the partners, or even passed on by inheritance. Sociétés de capitaux have anonymous partners holding freely transferable shares.

Finally, public companies are divided into three groups: industrial and commercial public establishments, state-owned companies and semi-public companies.

According to Stephane BALLAND and Anne-Marie BOUVIER (2008), the company is a multidimensional entity: economic, human and social[16] .

- **The company is an economic player**
- It is a production unit that transforms inputs into products and services;
- It contributes to gross domestic product (GDP) by generating added value;
- It is also a spending unit that consumes and invests to ensure production;
- It is a unit for distributing and sharing wealth;
- Because of its commercial nature, it is subject to constraints of effectiveness (achieving fixed objectives) and efficiency (achieving objectives while optimising resources).
- **The company is a human reality**
- The company is also defined as a community, a human group, employees who contribute to the achievement of common strategic objectives;
- Individuals must work together to achieve objectives;
- Company interests and individual interests must converge.
- **The company is a social reality**
- The company creates jobs, income and products, but also innovation and technological progress;
- The company acts on its environment, its activity having repercussions on those of other economic agents;
- Companies take certain actions spontaneously or under pressure from the environment.

Another term much used in this context is "entrepreneurship". Entrepreneurship is defined as the process by which an individual or group of individuals put time and capital into the search for market opportunities, with a view to generating value and growing the business through innovation, whatever the resources available. It's a process that involves launching a project, organising the necessary resources and assuming both the risks and the rewards.

---

[16] FAYAL Alan op.cit pp 25 - 35.

11.5.1.2. Definitions of the entrepreneur and entrepreneurship

But what about the "entrepreneur", a term that is often used in everyday life and particularly in the business world?

The entrepreneur is the person who sets up or runs the business. He or she possesses special qualities and implements innovative working methods.

Famous authors have given their take on the entrepreneur:

- For SCHUMPETER (1950), an entrepreneur is someone who is willing and able to transform an idea or invention into a successful innovation.
- For Peter DRUCKER (1970), an entrepreneur is a person who is prepared to put his career, his time and his capital at risk in order to realise an idea, sometimes even under risky conditions.

In the vocabulary used to describe a company director, we quite often recognise terms such as *courage, innovation, intuition, creativity, invention, perseverance, taking the initiative, risk-taking*, etc.

We will conclude this series of definitions with the notion of entrepreneurship. The Centre d'analyse des politiques économiques et sociales (CAPES) in Burkina Faso, in its December 2004 publication entitled *"Les fondements de l'entreprenariat au Burkina Faso"* defines entrepreneurship as "the creative ability of the individual, alone or within an organisation, to identify an opportunity and seize it to produce new value ...". In other words, it is the ability of a person or a group of people to embark on some kind of adventure to create something new, with all the risks that this can entail.

11.6. WORK PRIOR TO SETTING UP A BUSINESS

11.6.1. What you need to know beforehand

Before tackling the preliminary work involved in setting up a business, we should ask ourselves the following question: Why set up a business? This question can be answered as follows:

Setting up a business means :

- First and foremost, become an entrepreneur;
- The quest for financial freedom, success and practicality;
- To satisfy a desire to change the world;
- Coming out of a period of unemployment;
- Leaving an uncomfortable working environment;
- Building for the future.

But you need to be aware that setting up a business and, above all, running it involves taking risks, which include[17] :

[17] LUKENI LUNYIMI, Comment creer une PME formalite juridique essentielle, 2nd edition Kinshasa, 1997.

- Loss of money: the adventure of setting up a business can be short-lived, resulting in a total loss of the money you put into it.
- Waste of time: setting up and running a business requires a gift of time. It can even take a long time to think things through, to find your bearings and sometimes to go through a number of tedious procedures, especially when the administrative facilities are not in place.
- Losing friends or part of your family: setting up and running a business sometimes means being away from your friends and even some members of your family, which is unpleasant and difficult to bear.
- Loss of credibility: if you fail, you can lose credibility with the people around you.
- Loss of self-esteem and self-confidence: even when you fail, you can become so distressed that you doubt yourself and lose your self-esteem.

How do you manage these risks?

- Don't be afraid: fear of risk kills you; you need to be brave enough to overcome any difficulties.
- Form an idea and develop it over time: you need to set yourself a goal and give it enough thought day by day.
- Don't wait too long to make up your mind: you can't stay in a state of indecision for too long, otherwise the desire to set up your own business will eventually fade.
- Self-confidence: self-confidence is the key to success.
- Classify risks according to their importance: not all risks are equally serious. They need to be identified and prepared for accordingly.
- Classify risks according to their probability of occurrence: not all risks will necessarily occur or occur at the same time.
- Think about mitigation measures: depending on the nature of the risk, prepare the appropriate solution to deal with it.
- Taking action: as soon as the idea is ripe and the conditions are right, it's time to act.

It can be difficult to form an idea, especially when you're setting up your own business for the first time. There are some practical tips for forming an idea with a view to setting up your own business. The following is a non-exhaustive list:

- Focusing on a particular market: something that's very much in demand at the moment.
- Meet the players: suppliers, customers, service providers, etc.
- Try to understand their problems: what are they looking for? What do they need?
- Observe what is already being done: who is currently doing what?

- Strive to do better than what already exists: focus on quality, speed, simplicity, reliability, cost, etc.

Identifying an opportunity: is there a promising sector at the moment?

- Looking at what's being done elsewhere: what experience can we learn from the outside?
- Examine your own strengths and weaknesses: what can you do better than others and what can't you do better than others?
- Look for sources of inspiration: read specialist magazines, websites, talk to professionals, take part in competitions, etc.
- Defining a grid for analysing ideas: choosing precise criteria for analysing ideas.
- Identify the ideas and sift through them using the analysis grid: eliminate the least interesting ideas according to the criteria selected and concentrate on the ideas that offer hope.

But what can you expect when you become an entrepreneur? It's important to remember that being an entrepreneur will change the course of your life in a number of ways:

- Running a business is a demanding job;
- Once you're an entrepreneur, you have little free time for family and friends;
- The contractor often has to work outside office hours;
- He sometimes has little or no holiday;
- Setting up a business requires a lot of financial investment at the outset before the first profits start to flow in.

But none of this detracts from the nobility of the entrepreneurial profession, because all entrepreneurs are passionate about their work.

11.6.2. The key stages in setting up a business

Guilhem Bertholet (2015) in his book entitled **"Le petit livre rouge de la création d'entreprise"** has identified twelve key stages in business creation, which he calls **"Les douze premiers travaux pour la création d'entreprise".** We have reproduced them below with explanatory comments to make them easier to understand:

1. Understanding your personal goals:

a. Getting to know yourself;
b. Get to know the other members of your team;
c. Know your expectations and those of your team;
d. Find out whether your project is succeeding or failing;
e. Knowing when to stop the adventure or start a new one.

2. Defining your business objectives :

a. Formulate a mission statement that sums up the whole company in a few words

b. Examples:

i. For a landscaping company: *"Make the garden the most beautiful part of the house".*

ii. For a hairdressing salon: "*The most talented and friendly hairdresser in the area".*

iii. For a website comparing car dealers, garages and repairers: *"Find a garage you can trust".*

c. Formulate precise and concrete objectives, for example :

i. "To produce and market 1,500 tonnes of tomato concentrate in 2018.

ii. "To capture 25% of the national fruit and vegetable market by 2020.

3. Identify the problem you want to tackle:

a. The most important thing is the customer, and more specifically his needs;

b. What problem do you want to solve for the customer?

c. Example of a tutoring company:

i. Helping students achieve top marks;

ii. Reducing the risk of repeating a year;

iii. A guarantee of greater professional success later on.

4. Know your step:

a. Define customer types;

b. Understanding the specific needs of customers;

c. Understanding customer behaviour;

d. Know your competitors;

e. Understanding major market trends.

5. Defining your business model:

a. Approach to building a business model:

i. Strategic partners: who are they?

ii. Key activities: what key activities do we need to set up to produce and sell what we want to offer customers?

iii. Key resources: what resources are needed to produce and sell what we want to offer customers?

iv. Cost structure: what are the main costs associated with our model?

v. Customer relations: what kind of relationship do our customers want us to have with them?

vi. Distribution channels: how do we want to make our product available?

vii. Revenue streams: how do customers pay?

viii. Customer segments: who are our main customers?

b. Build your action plan :

a. List the first actions to be taken:

i. Company name ;
ii. Company logo ;
iii. Main elements of my home page on the website;
iv. List the main customers;
v. List the main competitors.
b. Draw up a schedule for implementing the various tasks.

7. Find your first customer:

i. A company is only as good as its customers;
ii. The sooner we get our first customer, the sooner we can get the next one;
iii. List all the practical ways you can get in touch with your first customer
iv. Telephone ;
v. . Electronic mail ;
vi. . Participation in a trade fair ;
vii. Taking part in a conference;
viii. Social networks: Google, Facebook, LinkedIn, Twitter, etc.

8. Building your image:

a. What customers remember is the company's image;
b. The image is made up of several elements:
i. The personality of the company: dynamic, quality, human-sized or family-run;
ii. Graphic elements: colours, logo, font, etc.
iii. Content: what you say when you present your company.

9. Meeting people, lots of people:

a. Above all, meet people who will be of great use to you:
i. Members of your family ;
ii. Friends ;
iii. Potential customers.
b. Meeting opportunities :
i. Local events ;
ii. Within associations ;
iii. In sports clubs ;
iv. On public transport.
c. Leave a little souvenir for each occasion: your business card, a leaflet about your company.

10. Define your offer :

a. The offer consists of products and/or services.
b. Before defining the offer, it is essential to :
i. Take the time to understand your needs;
ii. Identify the people for whom this need exists;

iii. Find out what's already being done.

11. You surround :

a. Faced with the multitude of decisions to be made and possible choices, the entrepreneur is sometimes alone.

b. You need to surround yourself with people you can trust to :

i. Helping you take a step back;

ii. You provide guidance and advice:

iii. Helping you get your head above water.

12. Create your own routines :

a. The entrepreneur should impose "routines" (ritual tasks) on himself:

i. Call five potential customers who have already been identified;

ii. Writing an article for your blog;

iii. Write an e-mail to everyone you've met since the start of your project;

iv. Read magazines about your sector of activity.

This advice is not a panacea. But if you apply them correctly, you'll avoid unnecessary mistakes and be on your way to realising your business idea.

11.6.3. FORMALITIES FOR SETTING UP A BUSINESS IN THE DR CONGO

For a long time, it was particularly difficult and time-consuming to set up a business in DR Congo, especially for young people. Nowadays, the Congolese government has introduced facilities that make it easier to set up a business, even in the space of a week. This is the mission of the specialised one-stop-shop structures under the supervision of the Ministry of Finance, following DR Congo's accession to OHADA in 2017. These centres welcome and support entrepreneurs in the process of setting up, modifying and taking over businesses. The business formalities service enables an entrepreneur to complete the formalities associated with[18] in a single place and in record time, on the basis of a single document:

- Entry in the Trade and Real Estate Credit Register (RCCM),
- Registration with the tax authorities,
- Social security membership,
- Trade registration.

These formalities are described below:

II.6.3.1. For individuals

1. Formalities to be completed :

- Trade and Personal Property Credit Register (RCCM) ;

[18] LUKENI LUNYIMI, Comment creer une PME formalite juridique essentielle, 2nd edition Kinshasa, 1997

- Declaration of tax existence and unique financial identifier (IFU) number ;
- Carte professionnelle de commerergant (CPC) ;
- Notification of employer (CNSS) ;

2. Documents required for all formalities :

- 1 legalized photocopy of the promoter's identity card or passport ;
- 1 extract from the promoter's criminal record less than three months old or a declaration of honour duly signed by the promoter;
- 1 copy of the marriage certificate (if applicable) ;
- 1 certificate of residence for the current year (payment of the residence tax to the Service des domaines, and issue of the certificate of residence at the town hall or police station;
- One of the following documents in the name of the company founder: 1 registered commercial lease, a title deed, an urban housing permit, a certificate of plot allocation, a water or electricity bill;
- 2 identity photos of the promoter ;
- 1 application form for a business card to be paid on the spot at the counter at a cost of 500 Francs;
- 1 location form stamped by the company's tax office.

11.6.3.2. For legal entities:

1. Formalities to be completed :

- Trade and Personal Property Credit Register (RCCM) ;
- Declaration of tax existence and unique financial identifier (IFU) number ;
- Employer notification (CNSS).

2. Required documents :

- 1 photocopy of the identity card or passport of the managing director(s) and one of the partners;
- 1 extract from the criminal record of the manager(s), less than three months old, or a declaration of honour form duly signed by the manager(s) (pre-established form available);
- 1 copy of the company's articles of association ;
- 1 copy of the minutes of the constituent general meeting ;
- 1 copy of the notarial deed of subscription and payment of the capital or the declaration of regularity or conformity ;
- One of the following documents in the company's name: 1 commercial lease registered with the tax authorities, a title deed, an urban housing permit, a certificate of plot allocation, a water bill;
- 4 copies of the MO form (pre-established form available) ;

- 2 deeds of deposit at least ;
- 1 location form stamped by the company's tax office.

1. The secret of success

According to a study by Smail Business Trends, 72% of SME managers feel overwhelmed by their roles and responsibilities. The world of entrepreneurship.

> Never lose sight of the structure's objective;
> Surround yourself with the right people;
> Separate the wheat from the chaff;
> Good = Able to produce the expected result
> Bad = unable to add value

This is the potential for creating added value in lean management concepts

2. The productivity factor

The population qualified for a given position can be broken down into 5 unsold groups

a) Performance (20%)
b) Efficient performers (30%)
c) Inefficient performers (17.5%)
d) Toxic people (2.5%)

3. The keys to becoming a super manager

1. Acting coherently
2. Approaching the right skills
3. Optimising time
4. Leading a team
5. Overcoming relationship difficulties
6. Developing your leadership
7. Managing a project (bonus)

a) Cost/Delivery/PERFORMANCE (Quality) triangle b) Setting up an ad hoc power structure c) A dedicated team

4. The difference between creating and inventing

The difference between creating and inventing is that 'to create' is to draw something out of nothing, to make something out of nothing, whereas 'to invent' is to conceive, to have the idea first, of something new with a practical use.

11.7. THE THEORIES OF MICHEAL PORTER

The six forces of competition The value chain The concepts developed by Michael PORTER are indispensable in strategic case studies. The first of these was presented in 1982, in the book Strategic Choices and Competition. Three years later, the book summarised here uses the value chain to address several fundamental questions about competitive advantage. How can sustainable competitive advantage be achieved? How do interconnections strengthen competitive advantage? What are the strategic implications of solutions to these two issues? The essence of our study[19] .

11.7.1. Gaining a competitive edge

Before getting to the heart of the matter, we need to define the principle of competitive advantage. This is what the author does, describing it as "the value that a firm can create for its customers over and above the costs incurred by the firm to create it". It is therefore vital for a company to identify its sources of competitive advantage, before of course exploiting them.

A. A tool for analysis: the value chain

This exploratory work is based almost entirely on the use of the value chain as an analytical tool. It maps out the interweaving of value-creating activities by distinguishing between core activities (internal logistics, production, external logistics, marketing and sales, services) and support activities (procurement, technological development, human resources management and the firm's infrastructure).

This breakdown shows the impact of each activity in terms of costs or its potential for differentiation.

On the other hand, these activities are linked to each other by optimisation mechanisms (it may be necessary to arbitrate between two activities) or coordination mechanisms whose impact on the firm's costs and performance is considerable.

There are also external (or vertical) links, where the firm's value chain is in contact with those of its customers, suppliers and distributors. It thus becomes part of a value system (according to the terminology used in the book).

The different sources of competitive advantage then become clear.

In one way or another, these changes are reflected either in a shift in the costs borne by the company, or in an impact on its differentiation.

These two types of competitive advantage, combined with the field of activity on which the firm relies to obtain them, define three basic strategies for achieving results that are above the sector average. These are cost domination, differentiation and business concentration. However, the latter is unusual in that it is based on exploiting a competitive advantage (whether through costs or differentiation) within a narrow target.

The author strongly emphasises the danger of refusing to choose between these three basic strategies. In his view, "sticking to the middle path" inevitably leads to results that are below the industry average, unless all competitors make the same mistake. This should not prevent a company from reducing costs that do not involve sacrificing differentiation, or from seizing differentiation

---

[19] Prof Bernadin M. Performance management, CEPROMAD, DEA, 2019.

opportunities that are not costly. Quite simply, beyond these adjustments, it must choose the advantage that will ultimately be its own.

B. The cost advantage

Such an advantage can only be obtained by carrying out value-creating activities at a cumulative cost that is lower than that of competitors. The value chain is therefore once again the preferred instrument for the author's analysis. It allows us to study the costs associated with the value-creating activities and not with the firm as a whole. It then becomes possible to associate costs and assets with these activities. The resulting comparison can reveal potential cost improvements.

However, it is the analysis of the behaviour of the costs of activities, and therefore of their factors of evolution, that should be the focus of attention here. These factors are (according to the author) ten in number: economies of scale, the learning effect, the configuration of capacity utilisation, links, interconnections, integration, timing, discretionary measures, location and institutional factors. These factors combine to determine the cost of each activity and therefore the firm's competitive position.

This work, carried out on the basis of static data, must be completed by a study of the dynamics of costs. In this case, the aim is to predict the direction of variation of the factors of evolution and therefore to identify the activities whose costs will increase or decrease.

By proceeding in this way, a company gives itself the means to determine its relative position with regard to costs. Even an approximate comparison with the situation of its competitors enables it to choose between gaining an advantage by controlling the factors driving cost trends and reshaping the value chain (by improving design, manufacturing, distribution, etc.). It is also possible to carry out these two actions simultaneously. A sustainable cost advantage can only come from a combination of such measures.

Finally, it should not be forgotten that the cost advantage only leads to above-average results if the firm offers acceptable value to the customer.

The differentiation achieved by a company is the value it creates for its customers by meeting all the purchasing criteria.

There are many sources of differentiation. They result not only from the attributes of the product or the commercial policy, but from all the activities in the value chain, and even from downstream activities. There should be no confusion with the notion of quality, which is only one component of differentiation.

The strengthening of differentiation results from the multiplication of the elements of uniqueness or singularity from which the firm benefits. In fact, the links between the company's value chain and that of the customer are as many opportunities for differentiation.

However, the value created in this way remains to be seen. The customer only pays for lost value. They may even pay a higher price for a lower value, if the latter is better signposted. The success of such a strategy therefore depends as much on signage criteria (advertising, reputation) as on usage criteria (value actually created: product quality, delivery times, etc.).

Differentiation produces above-average results when the value lost by the customer exceeds its cost. The latter is linked to the cost evolution factors in the activities that generate the firm's uniqueness. Performance will be all the more sustainable if customers constantly perceive the increased value and competitors cannot imitate it. We must, however, act with restraint and avoid excessive differentiation, or equate uniqueness with the value created.

D. Factors influencing both competitive advantage and sector **structure**

Technology is the first of these elements. This term covers not only research and development activities, but all the technologies used by the company, whatever their nature (for example, a set of procedures). The value chain will therefore once again serve as an analytical tool. It is all the more important where there are considerable interdependencies with the technologies of customers and suppliers. The distinction between 'high' and 'low' technologies is therefore irrelevant here. All that matters is the link between technology and competition[20]

.

While technology has a direct influence on costs or differentiation, it also plays a part in competitive advantage by modifying other factors in the evolution of costs or uniqueness. It can also influence each of the five forces of competition, particularly in terms of barriers to entry. It is therefore advisable to give priority to technologies that have the greatest lasting effects on costs or differentiation, which has nothing to do with their degree of sophistication.

Because of this major role in gaining a competitive advantage, we also need to consider the revolution in technology. By giving itself the means to anticipate it, a firm can take the appropriate initiatives and thus appropriate or reinforce a competitive advantage.

This forecasting work is usually based on the life cycle model. During the growth phase, innovation focuses mainly on the product. Once they reach maturity, the aim is to rationalise mass production, which is why efforts are focused on improving manufacturing processes. When the decline approaches, innovations become rare, as technological investments reach the threshold of diminishing returns.

---

[20] Op.Cit pp 10 - 15

However, it should not be forgotten that technology forecasts must be treated with caution, given the high degree of uncertainty in this area. This applies just as much to the choice of technologies to be developed as to the decision as to whether or not to be a pioneer, or to the granting of operating licences.

The choice of competitors is the second element to influence both competitive advantage and the structure of the sector. This is perhaps the most interesting chapter in the book, as it goes against many conventional ideas. The reasoning is that competitors can make the firm more competitive and improve the structure of the industry. It may therefore be preferable to deliberately forego an increase in market share. Things are not so simple, however, in that it is a question of how to behave towards "good" competitors, whereas it would be better to focus attacks on "bad" ones.

Michael Porter explains the benefits of having well-chosen competitors as follows.

Such a competitor can act as a shield for the company in several ways. By absorbing fluctuations in demand, it can maintain a high level of activity despite a deterioration in the economy. By serving uninteresting, unprofitable segments where customers have considerable negotiating power. By having higher costs, which enables the company to generate higher margins. By stimulating creative capacity through the basic phenomenon of competition...

On a more general level, the presence of competitors makes it possible to avoid proceedings for dominant position or even monopoly (the example of Microsoft immediately springs to mind). Above all, it acts as a deterrent to the entry of a new firm. It makes it more likely that violent retaliation will be unleashed against a new entrant. The latter may already have been discouraged by the mediocre situation of the "good" competitors, who are an illustration of the difficulties experienced by second-tier firms.

To achieve these effects, we first need to determine the characteristics of a good competitor. Schematically, he must be credible and coherent in his decision-making, but at the same time suffer from weaknesses of which he is aware. This limits his ambitions, and therefore the risk of his actions going against the firm's strategy, but leads him to adopt a strategy that reinforces the favourable elements of the sector's structure. Obviously, no competitor is 'good'.

(The same reasoning will have to be applied by a firm that is not in a position to become a leader. It will have to choose a sector controlled by a 'good' leader, i.e. a company whose strategy will enable it to benefit from protection behind which the firm can live and be profitable).

A firm wishing to move closer to such a competitive configuration may first pursue a policy of deterrence and collective retaliation, targeting "bad" potential competitors. Conversely, the entry of good competitors will be facilitated by the conclusion of supply or distribution agreements, or even the granting of licences to exploit a technology.

Whatever the means employed to achieve this, the ultimate goal in this area is to achieve a sufficient market share to discourage any attack and which, combined with the other competitive advantages, preserves the balance of the market.

E. Interactions with the competitive field

This part of the book examines the interaction between the competitive field and the advantage held by the firm in a sector. It looks at the ways in which a sector can be segmented and at the factors that replace a product.

The purpose of segmenting a sector is to determine the firm's competitive field and therefore the segments it must serve.

Differences between products give rise to segments if they modify the intensity of one of the five forces of competition or when they influence the conditions of a competitive advantage. The segmentation variables are: product variety, customer type, distribution channel and geographic location. By combining these variables, an overall segmentation of the sector can be established. Taken two by two, they lead to the establishment of segmentation matrices.

The next stage concerns competitive strategy, since it involves determining the attractiveness of each of the segments that have been defined. This is, of course, a function of the structural attractiveness of the segment (by measuring the five forces of competition), but also of its size, growth, the firm's position and interconnections (i.e. the links that exist between several segments where activities in the value chain can be pooled). This last point is absolutely crucial.

Indeed, strong interconnections encourage the development of a strategy that corresponds to a broad target. But pooling activities entails coordination costs, trade-offs (when the value chain is not optimal for serving all segments) and rigidity. It is therefore not systematically advantageous to serve all the segments of a sector. In this case, the decision will be based on a strategy of concentration by optimising the value chain to serve one or a few segments.

This strategy will be viable against wide-target competitors if the optimal value chain (adopted by the firm) differs significantly from that required to serve other segments. On the other hand, in the face of imitators, it will only work if the firm enjoys a sustainable competitive advantage as a result of economies of scale (even modest economies of scale in a small segment enable the firm to maintain its advantage, especially if they cannot be offset by interconnections).

Replacement consists of supplanting a product or a production method by performing one or more specific functions in its place. This definition is important in that it avoids making a methodological error in identifying replacement products. This is because the search for products with the same general functions (i.e. an identical role in the customer's value chain), and not the same form, is what needs to be done.

Furthermore, replacement products are not always different products. The customer may, for example, decide not to buy anything at all, but to forego the corresponding functions. Technical progress, for its part, often makes it possible to reduce the rate of use of the product. The use of used, recycled or reconditioned products should not be overlooked. Upstream integration (often through direct purchase from the manufacturer) is the last potential replacement method. However, the threat of replacement comes not only from replacement products, but also from downstream replacements directly affecting the customer. A product may be affected by the disappearance of a complementary product even though it is not directly threatened.

The replacement mechanism is the result of a combination of three factors. The first is the comparison between the value and price of the replacement product and that of the product in the sector. The author refers to this as the relative value to price ratio (RRVP). The second factor is represented by the conversion costs required to adopt the replacement product (adoption of new sources of supply, relearning, risk of failure, etc.). The third is the customer's propensity to change product (largely dependent on risk profile or previous replacements).

The evolution of the threat is therefore a function of changes in relative prices, in relative value, in the value lost by customers, in conversion costs and in the propensity to change product.

These changes determine the replacement path. This begins at a modest pace, in the information and trial phase, followed by the take-off phase "towards an upper limit constituted by the number of potentially interested customers, a number that can of course vary over time as a function of technological developments.

In view of the above, a company wishing to launch a replacement product can target its efforts on the customers most inclined to change product, improve its offering in areas where the RRVP is highest, reduce conversion costs, etc. Conversely, defending an existing product can involve discovering new uses for it that the replacement product has no effect on, redirecting the competition away from the strong points of the replacement product, or getting distributors to contribute to the defence.

There are many mistakes that can be made in the fight against replacement products, but the most serious is undoubtedly that of taking the product's maturity for granted and therefore making it impossible to replace.

11.7.1.2. Interconnections and horizontal strategy

The aim here is to describe the overall strategy of a diversified smoker. The central question is how to exploit the interconnections between the units to gain a competitive advantage.

A. Interconnections between business units

In the 1970s, many companies diversified, using synergies between their businesses as a pretext. Unfortunately, this policy did not pay off, no doubt due to a lack of discernment in acquisitions or the absence of the necessary analytical tools. The result has been a new trend: the decentralisation of activities.

However, the developments observed by Michael Porter led him to recommend the adoption of a horizontal strategy. This involves coordinating the firm's divisions in order to give it a comparative advantage. The author bases his reasoning on four points. Firstly, the 1980s saw a transformation in the mode of diversification: acquisitions concerned related fields. Secondly, as growth slowed significantly, priority was given to results, and consequently to competitive advantage. Secondly, technical advances are making it easier to exploit interconnections. Finally, only a horizontal strategy offers the global perspective needed to face up to multipolar competitors.

To arrive at these results, the author identifies three types of interconnection, which may coexist.

Tangible interconnections correspond to the sharing of activities between the firm's units. They can involve any value-creating activity and create a competitive advantage through lower costs or greater differentiation. They do, however, entail the costs of coordination, compromise and rigidity.

Intangible interconnections involve the transfer of know-how between the value chains of two units. Involving the diffusion of the same skills, they often lead to the standardisation of basic strategies. However, this is a delicate process, because know-how is a highly subjective concept.

Finally, competitive interconnections exist when a firm fights diversified rivals through several units.

B. Horizontal strategy

Diversified companies will find it difficult to achieve high overall performance simply by optimising the results of their various units. This is particularly true for companies where decision-making is largely decentralised, which is detrimental to the exploitation of interconnections. Indeed, the heads of the units formulate their strategies without the slightest consultation and may even move in incompatible directions. To solve this problem, an explicit horizontal strategy is needed.

This requires a multi-stage formulation. The first step is to identify the existing interconnections, to find any that exist outside the firm, to determine those with the competition and finally to assess their importance for competitive advantage. It then becomes possible to develop a coordinated horizontal strategy aimed at exploiting and strengthening the most important interconnections.

It is by proceeding in this way that a diversification strategy can increase a competitive advantage in sectors already invested in, or create a sustainable advantage in new sectors.

However, we must remain cautious when looking for interconnections. We must neither overlook them, nor imagine that the slightest superficial similarity in technology or procedures is a potential interconnection.

Furthermore, even interconnections that offer real benefits can prove difficult to implement. This is the case when the benefits provided are not (or do not appear to be) shared equally between the units. Unit managers may also fear a loss of autonomy, especially if the company's culture to date has been one of extensive decentralisation, with each division having its own identity.

To overcome these obstacles and resistances, the author proposes the implementation of a horizontal organisation. This links together the units within a vertical structure and thus facilitates interconnections. This organisation is based on four elements. The horizontal structure corresponds to a transverse division in certain areas, the grouping of units or partial centralisation. Horizontal systems involve cross-functional management of planning, control and investment decisions. Horizontal human resources practices are designed to facilitate cooperation. Finally, horizontal conflict resolution structures may be necessary. The combination of these horizontal elements with a vertical structure (though not a matrix structure) seems sufficiently innovative to the author to speak of a new form of organisation.

C. Complementary products

This is a special case of interconnection, where one product is used to complement others. Complementary products are therefore a form of link between sectors. Their existence requires a choice between three practices[19] .

Direct control of complementary products sees the company offer the full range of complementary products. It can strengthen a competitive advantage by taking advantage of interconnections or increased differentiation through a complete offering. Unfortunately, interconnections do not always exist, or some of the sectors involved may not be attractive. In any case, there are usually so many complementary products that the only solution is to focus on the most strategic ones.

Bundling is the sale of complementary products exclusively in blocks. This practice is sub-optimal, as it provides a single response to all customer needs.

Interconnections and increased differentiation can certainly make it interesting. However, there is still a risk that customers will be able to collect batches themselves from specialist firms offering the items on more favourable terms. This risk tends to increase as customers acquire mastery of the technology and therefore the ability to buy individual items.

Finally, in cross-subsidisation, the firm sells a basic product on terms that facilitate the sale of more profitable complementary products. This presupposes a high price sensitivity for the basic product, but a low one for the profitable good, as well as a close link between the two. Of course, there is a risk that the customer will only buy the basic product, while the conditions mentioned above may disappear as the sector evolves. In this case, the company must be prepared to abandon cross-subsidisation.

11.7.1.3. Strategic implications

The author concludes this book by developing the impact of the preceding reasoning on competitive strategy. He gives some ideas on how to deal with uncertainty, as well as on offensive and defensive strategies.

A. The role of sectoral scenarios

Until the 1980s, the scenarios developed by firms focused on macroeconomic and macro-political factors. However, such macro scenarios are not relevant to the analysis of a particular sector. This gave rise to the need for sectoral scenarios.

There are several stages in the process of developing a coherent vision of what the future might hold. First, the uncertainties likely to influence the structure of the sector are identified, and the causal factors are determined. The hypotheses relating to these factors are then combined to produce coherent scenarios. It then remains to analyse the implications of each of the scenarios (structure of the sector, sources of competitive advantage or behaviour of competitors). There are, of course, feedback loops between these different phases.

Once the scenarios have been worked out, the company has five options. It can bet on the most likely scenario or the one that is most favourable to it. It can also choose a strategy that is viable whatever the scenario (but such a compromise is never optimal), or one that allows it to retain sufficient flexibility until one of the scenarios materialises. The last solution is to use our resources to encourage one of the scenarios to materialise.

As these solutions are not mutually exclusive, it is sometimes useful to combine them or switch from one to the other.

B. Defensive strategies

A strategy aimed at defending against a new entrant is evolutionary. It corresponds to the degree of progress made by the newcomer in its plan to

conquer the market.
The first step is to carry out research to gain a better understanding of the future market. Then comes the actual entry phase, accompanied by investments designed to build up a viable position. Then comes the implementation of the long-term strategy. Finally, the post-entry phase involves investments designed to consolidate the company's position.
The more advanced the entrant's process, the higher the barriers to exit (because of the financial efforts involved) and the more costly the defensive strategy will be. The most effective defensive tactics will therefore be those that discourage attempts, rather than those that aim to drive out a company that is well advanced in its process. This can be achieved by reinforcing barriers to entry (by filling existing gaps in the range, by increasing conversion costs for customers or by increasing capital requirements, etc.) or by increasing the expectation of a response (by signalling a willingness to defend, by clearly showing possible obstacles, by accumulating resources). It is also possible to reduce the incentive to attack by reducing the sector's profits.
In any case, the best defensive strategy is to dissuade any attack. If this fails, we will certainly have to switch to a reaction strategy, but always with the aim of altering the new entrant's perception of the sector and its chances of success.

C. Offensive strategies

An offensive strategy must above all not consist of a policy of imitating the leader, but on the contrary be based on a sustainable competitive advantage (whether in terms of costs or differentiation).
On the other hand, it will need to be close to the leader in other value-creating activities. Otherwise, its competitive advantage will not be enough to offset that of the leader.
In practice, the attacker can operate in three ways.
By reshaping the value chain, it can carry out certain activities differently. This solution is viable if it proves difficult for the leader to imitate.
A redefinition of the field of competition will take the form of either a widening to exploit interconnections, or a narrowing to serve only a particular target with an optimised value chain.
These first two strategies have the advantage of often hindering the leader's response, forcing him to go against his usual strategy in order to defend himself.
Lastly, spending in excess is the simplest option, but also the riskiest.

11.8. INTERNATIONAL STANDARDISATION

11.8.1. Definitions

- ISO 9001 defines the criteria applicable to a quality management system. It is the only standard in the ISO 9000 family that can be used for certification (although this is not an obligation). Any organisation, large or small, whatever its field of activity, can use it.

- The ISO 9001 standard is international and general. It is a guide for the management and organisation of a company or organisation, without defining ready-made solutions. So everyone can adapt it to their own culture and best practice.
- ISO 9001 is a management system standard.

Its requirements cover a number of management principles:

- Customer focus,
- Leadership,
- Staff involvement,
- The process approach
- Improvement
- The evidence-based approach to decision-making,
- Mutually beneficial relationships with suppliers
- The results, ...
- ISO 9001 is a manual or guide that brings together all the best practices for managing a company and ensuring customer satisfaction.

11.8.2. The benefits of ISO 9001

Regardless of the type or size of the company (SME, large group) or its sector of activity (service company, bank, private or public body, etc.), any company can obtain certification. The requirements of the standard focus on the organisation, not its sector of activity.

11.8.3. Drawing up standards

Experts from public authorities, higher education, industry and each ISO member country make proposals for standards.

They are themselves selected by the ISO members. The latter approve or reject the experts' proposals for standards.

11.8.4. Contribution of ISO 9001 to the company

Management principles enable the company to guarantee customers uniform, high-quality products and services.

It's a way of radically changing the way we work and improving our working methods. It's a way of motivating staff and establishing a culture of improvement.

ISO 9001 allows you to see errors, malfunctions and problems within the company.

Periodically, companies will be encouraged to take a step back from their organisations and objectives, and to take stock.

Certification is recognised worldwide, making it easier to penetrate markets around the world.

The standard therefore brings confidence to the company and sends positive signals to customers, guaranteeing the quality of products and services.

In today's world, these 2 criteria have become very important for a company's survival. Another criterion added to that of respect for the environment, which is driving more and more customers to use products or services that tend to preserve it.

11.8.5. ISO certification

To obtain certification, a 3-year audit cycle must be completed.

In the first year, the company will undergo a full audit, and then in the following two years will send a file to the certifying body, which will decide whether or not to issue the famous sesame. The certificate is valid for 3 years.

1. Guaranteeing quality through ISO certification

ISO certification to guarantee compliance:

- A product ;
- A service ;
- An organisation ;
- A process.

A certified company is one that meets the regulatory requirements for ISO certification.

The International Organization for Standardization defines ISO certification as "a procedure by which a third party gives written assurance that a product, process or service conforms to requirements specified in a standard".

2. ISO certification and family of standards

ISO certifications are quality management standards. Here is a table summarising the different families of ISO certification:

Table 6: ISO standards certification

ISO family	Area of action
ISO 9000	Quality management systems and their terminology
ISO 9001	The requirements of quality management systems and the obligations of certified companies
ISO 9004	Quality management guidelines for continuous improvement
ISO 10011	Guidelines for carrying out quality control audits
ISO 14001	Guidelines for respecting the environment

11.9. INFORMATION ABOUT ISO CERTIFICATION

- The International Organization for Standardization does not issue ISO certification. Specialised bodies are responsible for certifying companies, the best known in France being COFRAC.
- ISO certification is valid for 3 years. To extend certification, companies

must undergo a quality control or quality audit.

- ISO certifications guarantee the processes used to produce a product or service, but not the product or service itself.

11.10. IMPORTANCE OF STANDARDISATION

The use of ISO 9001, ISO 14001 and ISO 18001 standards is essential to meet international requirements. Of course, it is also essential for the company to ensure that the global creation system does not suffer from injustice linked either to social classes or, even less so, to geographical injustice.

11.10.1. ISO 14001: Environmental analysis

Environmental analysis in the case of rural areas is based on balancing the issues faced by rural structures in comparison with their urban counterparts across the board.

11.10.2. ISO 18001: Social and working conditions

Under no circumstances can an organisation function without financial resources, and these resources come from the proportional contributions of its beneficiaries.

The payment of possible duties to workers and taxpayers must be renewed in social conditions capable of generating revenue. This requires :

- Profitable employment and reduced unemployment;
- Entrepreneurial skills (family income-generating activities) ;
- Productive investors.

In short, ISO 18001 is the standard that seeks to analyse the working and social conditions not only of beneficiaries but also of service providers, whose mistreatment affects the quality of their services to the detriment of beneficiaries (purchasing power, unemployment or other professional social situations in the face of the cost of living).

11.11. THE PERFORMANCE MANAGEMENT MODEL (Ball Ball)

Managers, like Philippe Lorino, quoted by Professor Bernadin Mbol. In the performance management seminar, they describe performance as "everything in the company that contributes to achieving strategic objectives".

In this sense, it would be reasonable to admit that company performance exists essentially to achieve programmatic objectives.

These objectives are based on financial solidarity and risk reduction in the workplace, with the aim of contributing to the fight against poverty.

- In this sense, to be preformant is to achieve strategic and multiple objectives (for example, the efficiency, relevance or effectiveness of the actions undertaken by a Cesarean mutual).

Gimbert (1980) places performance at the centre of the triangle: performance triangle

Figure 5: Performance triangle

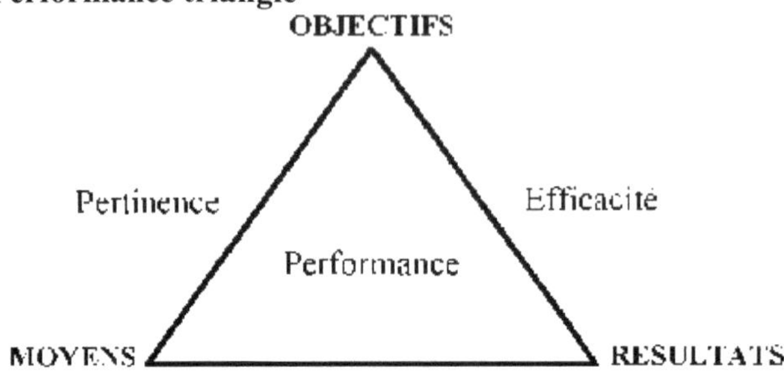

Efficiency

The segment between objectives and results defines **effectiveness**: Is the mutual sufficiently effective to achieve its objectives?

The segment between results and means defines **efficiency**: Does the mutual manage TO achieve its objectives at a lower cost?

The segment between means and objectives indicates **relevance**: Is the mutual providing itself with the right means to achieve its objectives?

This performance triangle is used by businesses and other types of organisation, and is regularly proven **in high-performance companies.**

However, today, this tool risks creating disturbances for the operator. In fact, this short-loop system does not place the person sufficiently in a place of human size and time. This distortion of the human dimension can lead to harmful effects such as socio-professional illness, depression, etc., all of which are a result of the person not finding their place in the system. Which is why ergonomics is not the issue here.

II.12. MANAGEMENT STYLES

We present here how autonomy can be developed through contextual management.

11.12.1 . Autonomous management triangle

Faced with more complex situations, employees' capacity for initiative and autonomy become crucial competitive assets. Managers play an essential role in developing autonomy by adapting their management style to their staff.

❖ **The performance triangle**

Autonomy levels are based on 3 criteria:

1. What the person knows how to do: their knowledge; their skills.
2. What it can do: its room for manoeuvre and the means at its disposal.
3. What she wants to do: her desire, her motivation

Figure 6: Autonomous management triangle

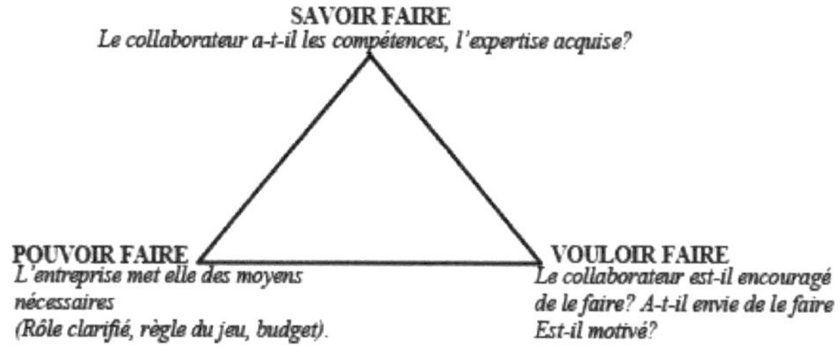

KNOW-HOW

Does the employee have the necessary skills and expertise?

POWER TO DO

Is the company providing the necessary resources?

(Role clarification, game rules, budget).

WANT TO DO

Is the employee encouraged to do it? Do they want to do it? Are they motivated?

11.12.2 Contextual management

Managing a team means managing people at all times, according to each person's specific tasks and in a changing environment.

The management compass allows you to choose the style for each employee according to their needs and degree of autonomy.

1. Very low battery life

The employee needs to 'know'. This may not be the case for a beginner or a newcomer who doesn't know the techniques or how to do things.

- The employee doesn't want to doesn't know
- The manager is directly involved here
- The management style is direct.

In essence, this will be a temporary style, invaluable in emergency situations.

2. Low autonomy

The employee begins to 'know' and 'have'. They develop their skills and are motivated by their work. The manager develops cooperation and can become personally involved by exchanging and confronting knowledge and ideas, values and convictions.

The management style is more persuasive.

It's a style that leaves room for initiative to the employee, while remaining firm on the objectives to be achieved. We share, we exchange, we discuss, with the major decisions remaining the responsibility of the manager.

Danger: overused, it's a stressful style. We can discuss the form, but it is utopian and illusory to discuss the cold - values, results to be achieved...

3. Average range

Employees have the capacity to innovate, because they are used to taking part in

decision-making.
The manager works on the axis of cooperation by disengaging.
The management style is associative.
It's a style that allows you to take part in decisions and make your own contribution.
The manager accepts decisions that he would not have taken himself, provided that the objectives are achieved.
Danger: the point here is to be a decision-making member, not to allow decisions to be made or to have decisions made that have already been made or that are of no interest.

4. Strong autonomy

Employees are already competent and autonomous: they have the means to make decisions themselves and already do. The manager delegates responsibilities and tasks.
It is discreet and provides follow-up and assistance on request.
The management style is organisational.
Meeting procedures are in place if necessary. Monitoring is ensured by dashboards.
Danger: there's no question of isolating yourself from the team or shirking your responsibilities.

5. Autonomy acquired but unstable

Employees have acquired autonomy, but when faced with an unstable system, they no longer know what to do. Change reveals the need for different skills and challenges existing ones.
Management needs to increase its level of involvement and participation.
It broadens the range of decision-making options by regulating. It provides flexible support.
The management style is negotiating in the face of the situation and the environment.
Here, the manager can use negotiation to provide support. Danger: once the change has been put in place, knowing how to go back to supporting employees in an organisational style.

11.13. ORGANISATION, COORDINATION AND MANAGEMENT OF THE PRODUCTION TEAM

As a crucial element in determining the selling price, it enables the transmission of instructions and rules to be applied to the members of the production team. The distribution, organisation and planning of work within the production team, the definition of training needs and support for training followed by the individual and collective performance of the team, leading and motivating the team and managing difficulties and conflicts.

In an organisation, teams are not really fixed objects. They are born, they develop in parallel with the company and they live a real history within the company. Therefore, the work on motivation and team cohesion is not negligible in the life of groups. Indeed, we are convinced that a team that works harmoniously together is infinitely more productive than one that does not.

11.13.1 What is a team

When talking about team management, the notion of team is very important and should be taken into account in all its aspects. In fact, we speak of a team when the absolute work is done together. This requires a number of elements, including complementary individual skills, interaction between the different members of the group and a clear desire to cooperate[21] .

It should also be noted that the team concept calls on a number of organisational rules and models of representation. It also takes into account a multi-disciplinary vision by learning to look at problems from several angles.

To produce quality work, a team must above all demonstrate a certain professional maturity. To achieve this, management training is needed to put in place a collective demand for quality. To measure a team's maturity, three criteria are often taken into account: cohesion, community and ethics.

11.13.2 Team management

The importance of motivation in team management Motivation is a very important concept that must always be brought to the fore. Good team motivation always multiplies performance, whereas the opposite wastes energy. It is therefore interesting to note that there are two essential principles in the functioning of a team:

Production and regulation

Those who produce are not necessarily those who regulate and vice versa. In each team, everyone is more or less willing to take on one or other function. Working as a group enables the members to perceive all the functional and even emotional problems and to get involved in solving them.

Team management is the key to creating a real group dynamic.

11.13.3 Role of the team manager

Whatever the size of the team or the department (sales, sales, marketing, etc.), to achieve the strategic objectives set by their management, managers need to buy the talent in their charge as fairly as possible.

They act as the link between management and collaborators. As well as having a genuine ability to understand the company's strategic challenges, they must also be able to convey this vision to their team, be able to synthesise ideas and

[21] Jacques - Daniel ROCHAT, Creer et gerer eme entreprise, Edition ROC Paris 2010.

communicate effectively.

Their primary role is also to know :

Acting as a bridge between top management and the various members of staff: ensuring that the company's values are respected, the strategy is followed, the messages are effectively conveyed, the budget is respected and the objectives are achieved, passing on information to the field, ensuring that staff are happy to be mobilised, proposing innovative actions, ideas and projects, ensuring that the objectives set are achieved both materially and in human terms: reviewing, integrating and orchestrating talent, giving staff the means to carry out the tasks required and ensuring that this is indeed the case;

Sharing a broader vision: conveying the company's values, explaining the reasons behind projects in their entirety (at the level of the team, the department or the company as a whole);

Giving meaning to the different missions: implementing an optimal dynamic and innovation, taking care of well-being at work, reducing tensions, conflicts and turnover...

Helping the team to progress: both individually (through personalised coaching, specific training or by encouraging autonomy and decision-making, etc.) and collectively, by encouraging collective intelligence and cooperation.

11.13.3.1 The keys to effective team management

To fulfil their role as effectively as possible, managers need to possess a range of skills and aptitudes, governed, once again, by common sense!

Organising, being able to do things, managers need to know how to organise their work and their time in the best possible way. They need to be able to orchestrate tasks and plan ahead, but also anticipate and manage unforeseen events (an employee's absence or departure, logistical or material problems, delays in deliveries, untimely interruptions, market tensions, etc.), which can bring an entire team to its knees if they are handled inadequately or completely ignored.

11.13.3.2 Communication

Communication is the basis of any healthy relationship. Managers need to formulate their demands effectively so that their staff carry out the tasks and achieve the objectives set, but they also need to allow their staff to express themselves freely, to put forward their ideas, needs, proposals, thoughts, feelings and so on.

11.13.3.3 Managers need to know how to :

- Listen fully: if you want to be heard and get your message across effectively, the first thing you have to do is listen.
- Giving and receiving feedback on a regular basis: taking stock, making

adjustments if necessary, enabling everyone to improve on a daily basis. Constructive criticism enables everyone-managers and employees alike-to move forward.

11.13.3.4 The right framework

It would be a mistake to have a linear management style, whatever the situation. Managers need to be able to adapt their management style to suit the context, the people they are dealing with, etc., in order to stay on course.

Different leadership styles are available:

> Direct: based on a mode that gives the manager maximum po- vis. Very little - if any - room for manoeuvre is given to employees. Can be interesting in times of crisis, for example, but generates a great deal of ill-being in teams.

> Persuasive: strong involvement of the manager in decision-making, while maintaining a human dimension in management.

> Delegative: a great deal of leeway is given to employees, who are regularly consulted for advice and decision-making, and are heavily involved in the life of the team and the organisation. The objectives are always very effective, and the manager must delegate the right task to the right person.

> Participative: the most open and the most humane: employees are heavily involved in the life of the team, particularly when it comes to cross-functional decision-making.

11.13.3.5 Motivate

Managers need to keep their staff as motivated as possible and be able to spot any signs of motivation as soon as they appear, so that they can get things back on track as quickly and effectively as possible. They must bear in mind that the sources of motivation vary from one individual to another, from one moment to another. Some employees need relatively confidential daily support and recognition in order to feel reassured and to progress, others appreciate that their work is recognised in front of the whole team, and still others are motivated by challenges, bonuses, and so on.

Motivation has different aspects:

- ❖ Sharing your vision with enthusiasm
- ❖ Setting an example, being real
- ❖ Giving meaning
- ❖ Encourage, value, recognise and reward effort, hard work and success...
- ❖ Involving people, giving them responsibility, empowering them, etc.

11.13.3.6 Enhance

The manager encourages, values and recognises the work, efforts, talents and skills of each individual. They enable everyone to develop and improve. They encourage collaborative working to give greater meaning to individual and collective missions[22] .

11.13.3.7 Being authentic

By being true to themselves, retaining their personality and respecting the company's values and practices, the manger forges links with team members and fosters good relations between his or her colleagues in order to establish a climate of mutual trust. He or she is not a superman or a robot, but someone who is capable of admitting that he or she doesn't know everything and can make mistakes if necessary. They must also be able to show indigence, tolerance and empathy towards their colleagues.

11.13.3.8 Strengthening the team

Projects will only see the light of day if the team is united, if everyone does their job while supporting their colleagues where necessary. Because it's a well-known fact: "alone we go faster, together we go further". The role of the manager is therefore to develop team cohesion and cooperation.

To improve team cohesion, there are tools such as team building. Used properly, they help to bind a group together and strengthen team spirit and collective motivation.

11.13.3.9 Inspire

The manager must be a driving force for the team, providing leadership, setting an example and involving all his colleagues in his dynamic. Benevolence, respect, authority and flexibility are essential on a daily basis.

Leading by example does not mean being untouchable and always right, possessing an infallible science. Managers need to be able to listen to criticism, question themselves and admit when they are wrong.

11.13.3.10 Saying no

The manager must know how to frame, reframe and, if necessary, say no firmly. Respect isn't earned by always saying yes, but by being able to refuse politely but firmly when necessary.

Saying no is a sign of respect not only for yourself, but also for others. As a manager, saying no shows self-confidence and asserts your position as a leader.

11.13.3.11 Managing conflict and change

These are two exercises that can't be improvised and that every manager will encounter at least once in their career. They need to be able to spot any signs of tension within their team, decipher and analyse the causes and act appropriately before the situation escalates and gets out of hand. Furthermore, change of any kind is a process that requires support. The phases an individual goes through

---

[22] Prof Angel ONSIN N'saman, Leadership, as a function of organisation and command CEPROMAD, DEA, 2020 pp 39 - 78

when a change is announced are immutable, and some can be unpleasant. Depending on the personality and experience of the employee, the context and the extent of the change, the manager will need to be cautious and adapt his or her management style and support.

11.13.3.12 Tools to manage

- Time and priority management
- Reperting techniques
- Running meetings
- Interviews (motivational, professional, recruitment)

The production team leader supervises the running of production operations, allocates the necessary resources according to requirements and checks the results.

Applying tools sparingly. He checks product quality, compliance with procedures (standards, rules of traceability, quality, hygiene, safety and respect for the environment) and customer specifications.

II PART

ECONOMIC MANAGEMENT MODEL

CHAPTER III. CONCEPT OF ENERGY

111.1.CONCEPT OF ELECTRICAL ENERGY

Electrical energy is the most important of all energies. From its production to its use, there is a whole chain of conversion, electromechanical, electromagnetic and electrical, to consider.

Although the concept of energy is omnipresent, even in everyday life, it is proving very difficult to define precisely.

❖ **Definition**

The energy of a system represents its capacity to modify, through interaction, the state of another system.

Energy, an abstract concept, comes from the Greek word "ENERGIA" which means force in action. In other words, the ability to produce movement.

The concept of energy is not easy to define, because energy is not lost as such. To put it simply, energy is what is needed to make systems work, i.e. to make them do work in order to produce an effort.

Energy comes in many forms: kinetic, chemical, solar, wind, water, electric, nuclear, mechanical, potential and thermal. Thanks to the input of energy, systems can produce visible or invisible radiation (e.g. torches, radios, etc.) heat (e.g. hotplates, water heaters, etc.) and set themselves in motion (e.g. alternators, turbines, etc.).

The Larousse dictionary gives the following definition of energy:

A quantity that characterises a physical system, retaining the same value throughout all the system's internal transformations (law of conservation) and expressing its ability to modify the state of other systems with which it interacts.

Energy is a fundamental element of our modern society: it is produced, transformed and stored.

There are many analogies, except that our energy manipulations can seriously disrupt our environment, because our demands in terms of transport and comfort mean that our energy requirements are growing at an inordinate rate.

Since the dawn of mankind, we have been burning fuels: first wood, then fossils (coal, oil, gas), and finally uranium. Just over a century ago, electricity, a modern form of energy, came to the fore.

111.2.ENERGY TRANSFORMATIONS

Energy is therefore a great measurable quantity, which manifests itself in the course of its transformations, i.e. in the course of its passage from one form to another.

Example: In a torch, the chemical energy delivered by the battery is transferred to the filament of the bulb via the flow of an electric current. This causes the bulb to heat up and incandesce.

Energy can be transferred from 4 fagons:

- Via electrical work - circulation of an electric current (previous example),
- Via mechanical work,
- Via heat,
- Via radiation.

Energy undergoes a chain of transformation when it passes from one system to another, or when it is modified within the same system.

For example: In our modern world, we consume a lot of internal energy in chemical reactions (combustion) or nuclear reactions. We convert this energy into mechanical energy in heat engines. We use this energy in this form or, after transformation, in the form of electrical energy.

111.3. PRINCIPLE OF CONSERVATION OF ENERGY

Any increase (or decrease) in the energy of one system is accompanied by an equal decrease (or increase) in the energy of other systems. There is no such thing as spontaneous energy creation.

Energy Sources

An energy source refers to all the phenomena from which energy can be extracted. These sources may be natural (referred to as primary energy sources) or artificial.

In the latter case, we talk about secondary sources of energy.

Primary energy is the energy available in the environment that can be directly exploited without transformation. Given the energy losses at each stage of transformation, storage and transport, the quantity of primary energy is always greater than the final energy available.

There are many sources of primary energy:

- Crude oil;
- Natural gas ;
- Solid fuels (coal, biomass) ;
- Solar radiation ;
- Wind power;
- Geothermal energy.

So it's essentially thermal energy and mechanical energy.

Thus, the mechanical energy produced by a windmill is primary energy. On the other hand, if this mechanical energy is converted into electricity, the electrical energy produced is considered as electricity, the secondary energy since it is obtained by transformation.

These energy sources may be renewable or non-renewable.

Non-renewable energy sources are energy sources that will disappear one day because their stocks on earth are limited. These are fossil and fissile energies.

Fossil energies are those derived from the composition of organic matter, mainly plant matter, over millions of years.

These include coal, peat, lignite, hard coal, oil and natural gas.

Renewable energy sources depend on elements that nature constantly renews. They are considered to be inexhaustible: the sun, water, wind, heat, wood, etc.

111.4. ELECTRIC ENERGY

Electric energy is the cleanest of all energies. From production to use, there is a whole conversion chain, electromechanical, electromechanical and electrical, to consider.

Electricity is a physical phenomenon due to the different electrical charges in matter, expressed as energy.

Electricity is a natural part of our environment, but it was only during the 19th century[e] that its properties began to be understood.

Lightning was the first visible manifestation of electricity for humans.

111.5. THE ROLE OF ELECTRICAL ENERGY

The role of electrical energy in the economic development of concepts no longer needs to be proven. Since the industrial revolution of 17802, based on the use of new sources of energy, including electricity, and considered to be the second revolution in the world after that of agriculture or the Neolithic, electricity has been changing all the habits of human activity.

According to the economic literature, a country's economic growth is often linked to the new availability in quantity and quality of electrical energy, which can provide new employment opportunities, more efficient investment in production equipment incorporating new innovations, advances in health, education and access to new information and communication technologies, and consequently an increase in economic growth.

For Rosenberg (1998), electrification in industry makes it possible to intensify production by automating it, which has the effect of improving the productivity of forms. Economic growth can therefore lead to an improvement in well-being and new increases in income, which encourage the population to acquire electrical appliances in the long term. So, a simultaneous increase in the population and in the number of electrical appliances can lead to a high demand for electrical energy.

Africa is considered to be the least electrified region in the world. Wolde Rural (2006) estimates the electrification rate in sub-Saharan Africa at just 26%.

According to Reinikka and Svenson (2002), insufficient production or poor quality infrastructure also affects the reputation of companies unable to deliver their orders on time. Thus, the inadequacy of the electricity supply discourages potential investors in electricity-dependent production activities and limits local industrial development.

Over the last twenty years, the energy sector in the DRC has experienced a number of crises of varying degrees of severity, linked to the theft of cables, illegal connections, drought, aggravated by the socio-political crisis of 1998, and long pauses in investment, resulting in huge losses of energy due to obsolete equipment.

SECTION I. DISTRIBUTION OF ELECTRICAL ENERGY

Before setting about determining the profitability of this project, it is essential to have a basic understanding of energy.

I.1 ELECTRICITY NETWORK

1.1.1. Definition

An electricity network is a set of information systems used to transmit electrical energy from production centres to electricity consumers.

Structure of an electricity network

Figure 7. Functional structure of a 220 V/380 V network from generation to distribution

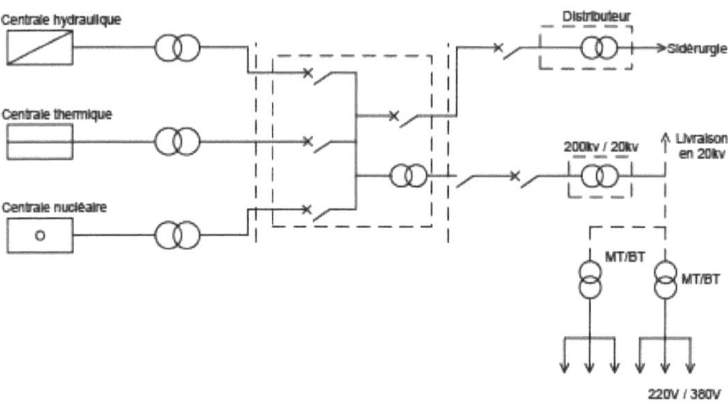

1.1.2. Types of electrical networks

Electrical networks are classified according to the following criteria

- The function to be performed ;
- The structure ;
- Operating voltage [24]

1.1.3. Depending on the function a. The transport network

The transmission network is high voltage (HV), from 70 kv to 220 kv, and its purpose is to transport energy from the major production centres to the electricity-consuming regions.

[24] Wikipedia encyclopedie libre http//w.w.w.wikipedia.com/ reseau electrique.html.page.consultee, le 05/9/2021

b. The interconnection network

The interconnection network is a network to which two or more sources are connected.

c. The distribution network

This is a network on which electrical energy is distributed at the same voltage level.

d. The distribution network

The distribution network is the network used to distribute electrical energy to different consumers.

I.1.4. According to the structure

The structure of the network gives us an idea of how it will be used and the various back-up options in the event of a fault.

We have :

A. Radial or antenna network structure

The radial network is the simplest form of network. The lines develop as antennae from the transformer station, at each node of the network. They can be connected to consumers. All non-ids are supplied by a single line. The cost of protecting this type of network is minimal, thanks to its simple structure. Security, on the other hand, is rudimentary, since a fault on one line and the opening of the relevant circuit-breaker leads to an interruption of supply for all downstream users. But its advantage is that it costs less. It is used for LV in areas with low load density. Although the security of supply is lower than that of the mesh structure, it remains high.

Figure 8: Schematic diagram of a simple radial network

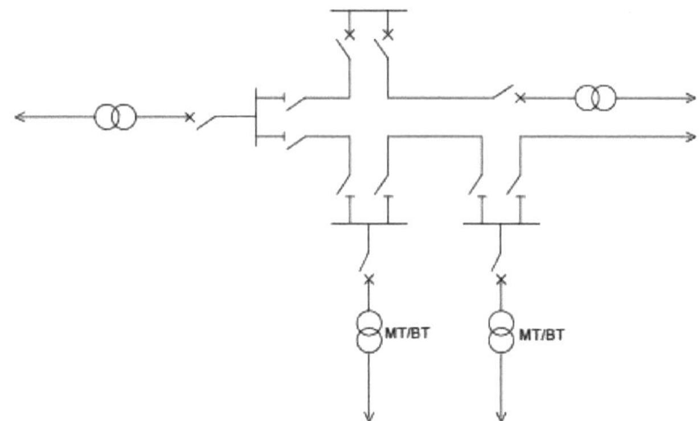

3. Mesh structure network

A mesh network is a network where all the line trunks are part of a loop; the graph of such a network looks like a spider's web or a net, which is generally very irregular. All the knots are fed from at least two sides.

There are two types of mesh networks:

a. Nodal-loaded mesh networks

Where all the loads served are concentrated at the network node. They are used above all for the high-voltage transmission network.

b. Networks with distributed loads

This is a network where the loads are connected along the lines; they are most commonly used for low-voltage distribution networks. Mesh networks provide maximum security, as any damage to a line trunk, leading to it being de-energised, will only affect the loads connected to the trunk.

c. Loop structure network

This type of supply is the distribution system most commonly used in urban areas. As the users are fed by a loop, the energy can arrive via two different paths, so switching off a line trunk on one of the paths, whether deliberately or not, does not interrupt their supply.

The criteria linking two sources must be able to withstand permanent overloads in the event of one of the sources being taken out of service.[23]

Figure 9. Representation of an open loop diagram

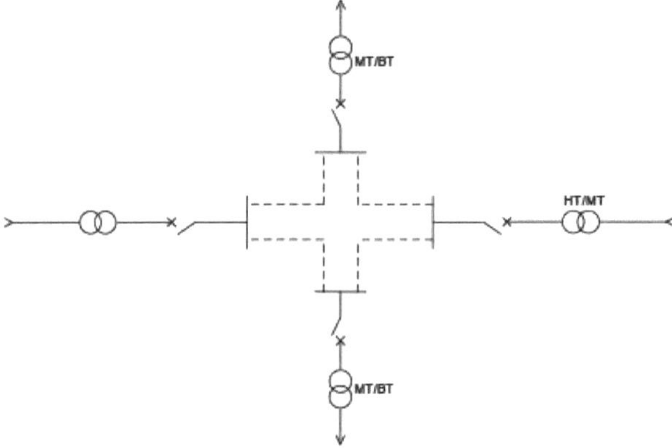

d. Tree structure network

This is the most commonly used structure for the lowest voltage level, i.e. low-voltage distribution.

There is little security of supply, since a fault on the line or substation will cut off all downstream customers.

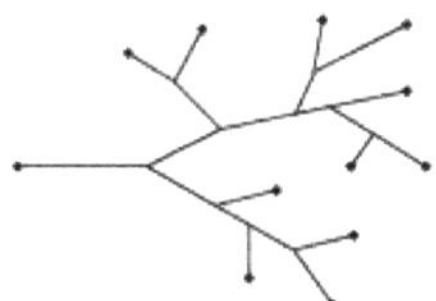

Figure 10 : Tree structure network

2. Depending on operating voltage

The table below will help us to understand the categorisation of voltages based on their values.

Table 7. Network categorisation by voltage

Designation	**In acronym**	**Value of lussions**	
		Encouraging alternative	**Direct current**
Very low voltage	T.B.T	From 0 to 50 V	From 0 to 120 V
Voltage base	B.T.A	From 50 to 500 V	From 120 to 750 V
Voltage base	B.T.B	More than 500 to 1000 V	From 750 to 1500 V
Medium voltage	M.T	From 1000 to 3000 V	From 1500 to 7500 V
High voltage	H.T	From 30 KV to 220 KV	More
Very high voltage	H.T.T	@220 KV	

SECTION II. BT DISTRIBUTION NETWORK

Understanding the distribution network will make it easier for us to implement the project.

11.1 Goal

The purpose of distribution networks is to supply the entire consumer, or an electricity distribution network is the part of an electricity network serving consumers.[25]

11.2 Characteristics of a LV distribution network

This network is essentially characterised by the way in which distribution is carried out and the level of voltage admitted.

BT distribution can be carried out in two ways

- The single-phase system: it has two wires, phase and neutral in the voltage is 220V.
- The three-phase system: it has four wires including three phase wires and a neutral wire whose voltage between phase is 380V and the voltage between

[24] DIANKABU et NDJIBU Critical study and proposal for improving the 1'unikin medium voltage network

[25] Wikipedia free encyclopedia.op.cit

phase and neutral 220V.

The low-voltage distribution network is a network to which domestic users are connected. The voltage used in. This network is classified as follows:

- From 50 V to 500 V for the low voltage network A
- Over 500V to 1000V for low-voltage LV networks

II.3. LV distribution network structure

Most of the low-voltage distribution network is operated as an antenna and generally in a tree structure.

Some networks in major cities are operated to ensure better quality of service.

Figure 11: LV distribution network

11.4. Construction of the LV distribution network

The low-voltage distribution network consists mainly of

by :

S Electric cabins ;

S Power lines ;

S Switching station.

11.4.1. Electric cabins

❖ **Definition of a MV/LV cab**

A distribution cubicle is a set of electromechanical equipment that protects and transforms electrical energy to supply low-voltage customers from the LV feeder.

11.4.2Cabin types

The cabins are grouped into two large families:

4- Outdoor cabins ;

4 Interior cabins

Within these two main families, MV/LV cabins can be divided into two categories:

- **The masonry booth**: This is a classic indoor type booth in which all the equipment is installed in a brick or concrete room.
- **Compact cab**: this is a cab in which all the equipment is housed in a metal box.
- **The open-air cabin (bottom of pole, top of pole)**: this is an outdoor type cabin in which all the cabin components, arranged separately in protective boxes, are isolated and exposed to the open air.

From the food point of view we have :

- **Antenna-fed cabins:** these are fed by a single derivation, with a single cable from the substation feeding the cabin; any intervention on this cable will cause the cabin's power supply to be interrupted.
- **Simple loop-through feeder cubicles:** these are fed by two cables from a single substation, and the two feeder cables together form a loop.

This system makes it possible to isolate the 1ere connection for maintenance purposes, while limiting the need to supply subscribers via 2eme connections.

For this power supply, it is only the failure of the source that cuts off power to the cabin.

- **Booths with double loop arterial cut-off:** This type of power supply is known as double derivation and is a distribution system that offers great continuity of service. The cab is connected to two or more cables from different sources, one of which normally supplies the cab, the others being reserved for recharging the cab in the event of a fault in the normal cab supply.

II.4.3. Methods of connecting cabins

The cabins are connected as shown below.

a) Arterial break connection

The cabins are connected in series on the network via the busbar.

The continuity of the framework is ensured by the busbars of the cabins it feeds.

Figure 12: Diagram of the arterial break connection

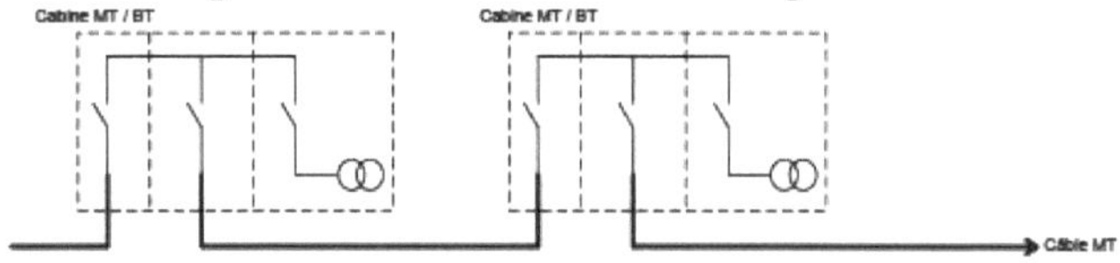

b) Double derivation connection

The cabins are served by two cables laid in parallel, one for work and the other for rescue.

The cabs are equipped with two interrupter switches and an undervoltage switch that automatically switches the power supply (work) to the emergency power supply in the event of a fault on the work supply.

Figure 13: Diagram of the double branch connection

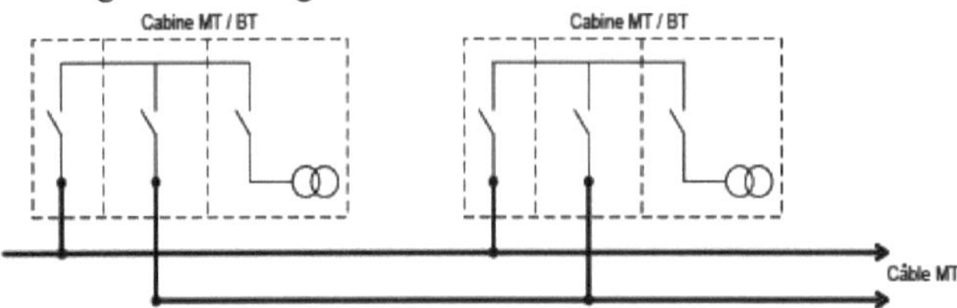

c) Antenna connection

With the antenna connection principle, a cabin can only be re-supplied from an external source following an incident or work on the network element (the power supply only).

11.4. Setting up a MV/LV cubicle

A MV/LV cab consists of three parts:

- Civil engineering;
- The medium voltage part ;
- The low-voltage section.

a. Civil engineering

The civil engineering part of a cab includes the construction of

Building the house ;

The medium-voltage part of a cubicle ;

- **The incoming feeder cubicle**: its role is to ensure the supply or separation between the distribution substation and the HV/MV network. It is equipped with a terminal box and a cable-head disconnector for an antenna cubicle, two cable-head disconnectors and a loop disconnector for cubicles connected in a loop. The incoming cell disconnector is always closed. There is also an isolation isolating switch in the loop cabins.
- **The loopback cubicle**: This cubicle contains: the shunt-mounted isolating switch for looping back the network if necessary. Here the isolators are open, and the two cells can be supplied from different sources.
- **The protection cell:** This cell is equipped with :
- Disconnector on circuit-breaker or load break disconnector with high breaking capacity MV fuse (rupto fuse) ;
- Magnetothermal protection relay ;
- Medium-voltage circuit-breaker.

The measuring and counting cell: this is made up of the following measuring equipment:

- **Amperemeters:** used to measure the intensity of an electric current;
- **Voltmeters:** used to measure voltage or potential;

- **Wattmeters:** used to measure active power;
- **Var metres:** used to measure reactive power;
- **Energy meter**: measures the amount of energy consumed.

Metering equipment: active and reactive energy meters, as well as potential and current transformers (TP.TI).

- **Transformer cubicle:** this cubicle contains one or more step-down transformers. If the cubicle is made up of several transformers, protection must be provided for each transformer.
- **Busbars:** busbars are rectangular or circular in shape, often made of electrolytic copper, and are connected to the R.S. and T. phases respectively. The approximate distance between busbars for a voltage of 20 kV is 21 cm, and for a voltage of 6.6 kV the maximum distances are 18 cm and 11.5 cm.

b. The low-voltage section

Here we find a low-voltage main switchboard (TGBT) with outgoing feeders to the subscribers, and a low-voltage circuit-breaker attached to the TGBT to protect the transformer against certain faults that may occur. The rated current of the low-voltage section depends on the load.

To sum up, in a cabin we have the following electromechanical equipment:

1. **Incoming cable head disconnector (STC) ;**
2. **Loop disconnector ;**
3. **Busbar disconnector ;**
4. **Low-voltage busbars ;**
5. **Circuit breaker disconnector ;**
6. **Protective relay ;**
7. **MV circuit breaker ;**
8. **Voltage and current transformers (TP and TI) ;**
9. **MV/LV transformer ;**
10. **Accessories: earthing switch**.

a. * **Potential transformer (PT)**: used to supply measurement equipment at a reduced voltage.

b. * **Current transformer (CT):** Used to supply measuring equipment with a reduced current.

c. * **Isolating switches**: these are devices which allow an electrical circuit to be opened or closed when there is no current; they have no arc-breaking device, and must therefore only be operated when there is no current.

They are used to visualise the state of an electrical circuit.

NB: there are also disconnectors that can open the circuit on load, they have arc extinguishing devices, they are called "on-load disconnectors" or high breaking capacity disconnectors.

d. **The cable-head disconnector**: used to receive the incoming voltage on the line.

e. **The earthing switch**: used to earth the line during an operating interconnection.

f. **The busbar isolating switch**: used to switch the voltage from one busbar to another.

g. **The busbar**: used to connect the incoming and outgoing power installations. They are made of copper or aluminium.

h. **The circuit-breaker**: this is an electromagnetic, or even electronic, protection device, whose function is to interrupt the electrical current in the event of an incident on an electrical circuit.

Figure 14: Low-voltage circuit breaker (with control unit)

i. **Lightning arresters:** protect all electrical installations against surges.

j. **The transformer:** This is a device which transforms a system of variable voltages and currents into a system of variable voltages and currents without changing their nature, but with the same frequency. Energy transfer

Figure 15: Step-down transformer

II.4.5. The code used in the LV circuit breaker network

II.4.5.1. Definition

By definition, electrical cables are equipment used to transport electrical energy from the distribution substation to the consumer.

a) General characteristics

1. Souls

They must meet the following conditions:

- Good conductivity: to reduce losses during energy transport, hence the choice :

- Copper: p = 18.51m .mm^2 /m at 20° c ;
    - Or aluminium: p = 29.41m .mm^2 /m at 20° c.

- Sufficient mechanical strength to prevent the conductor from breaking under load during installation, fixing and tightening of connections;
- Good flexibility: to facilitate the passage of conductors in ducts, to respect the trace of pipes, to supply power to mobile appliances ;
- Good resistance to corrosion caused by atmospheric agents and chemical environments.
- Good reliability of connections thanks to good resistance to the physico-chemical effects of contacts.

Souls can be

- In annealed copper, bare or with a metallic coating;
- Aluminium or aluminium alloy, bare or coated with a metal layer
- Or plated aluminium

Copper aluminium equivalence

For the same electrical resistance:

$$\frac{\text{section aluminium}}{\text{section cuivre}} = \frac{pAL}{pCU} = \frac{29,41}{18,51} = 1,59$$

This translates into the choice of an aluminium ame section immediately greater than that of a copper conductor in the normal range of conductor cross-sections.

Table 8: Copper aluminium equivalence

Section cui(mm)[2]

Section cui(mm^2)	1,5	2,5	4	6	10	16	25	35	50	70	95	120	150	185
Section alu (mm^2	2,5	4	6	10	16	25	35	50	70	95	120	150	185	240

Aluminium section (mm[2]

b) Section of cable used in BT25

Per phase for copper conductor

Current in A	Conductor cross-section in mm[26 27]

25SNEL. Etude et suivi d'd'dlectrification transfrontaliere et rural, Formation Kinshasa 05 - 08 juillet 2010

27Ditto

0 to 5A	1.5 mm^2
5 to 10 A	2.5 mm^2
10 to 20 A	4mm^2
20 to 25 A	6mm^2
25 a 32A	10mm^2
32 a 40	16mm^2
40 a 70 A	25mm^2
70 to 100 A	35mm^2
100 a 125	50 mm^2
125 to 160 A	70mm^2
180 to 200 A	95 mm^2
200 to 250 A	120 mm^2
250 to 320 A	185 mm^2
320 to 400 A	300 mm^2
400 to 500 A	2 150 mm^2
500 to 630 A	2 185 mm^2
630 to 800 A	3 185 mm^2
800 to 1000 A	3 240 mm^2
1000 to 1250 A	3,300 mm^2

c) Insulating jacket

This insulating jacket must ensure good insulation of the conductive core and have the following characteristics:

- Generals of all good insulation :
- High resilience;
- Very good electrical rigidity;
- Low dielectric losses
- Particular to the use of conductors and cables
- Good resistance to ageing ;
- Good resistance to cold, heat and fire;
- Vibration and shock resistance ;

II.4.5.2 Overhead power line

Overhead power lines provide a means of transporting electrical energy at reasonable cost, with relatively low losses and easier maintenance.

An overhead power line is essentially made up of :

- Supports or posts ;
- Drivers ;
- Accessories.

a) Supports or posts

The support for a line is a mechanical structure that can be :

- Metallic ;
- Wood;
- Concrete.

The main advantage of **metal supports** is their good mechanical strength, but they also have the disadvantage of being able to conduct electric current, so care must be taken to ensure that the insulators are in good condition.

Wooden supports have the advantage of not being good conductors of electricity, but they also have a major disadvantage in that they are often attacked by insects. They therefore require special precautions, such as the use of insecticides. One disadvantage of concrete supports is their weight.

b) Drivers

By definition, a conductor is a material that allows electrical current to flow easily. They must ensure good electrical continuity and be able to withstand external stresses without deterioration or breakage.

The constraints are :

c) Electrical constraints

- Voltage ;
- The intensity of the electric current ;
- Short-circuit current.[28]

d) Mechanical constraints

- The wind ;
- Frost - snow.

When choosing a conductor, we take into account :

f Voltage drop ;

f Warming up ;

S The economic section of the son ;

f The mechanical resistance of the wire.

There are several types of conductors, depending on their nature, such as :

- Copper ;
- Steel ;
- Aluminium ;
- L'almelec etc.

The most commonly used materials are copper and aluminium for cables, while aluminium cable is used for overhead distribution lines.

e) Accessories

The accessories are :

26LUZOLO.E, *electrical technology course notes*, 2ème gradual, ISPT - KIN, 2010 - 2011, page12

a. Insulators: these insulate the conductor from the metal part of the support. They are made of glass or porcelain, and there are rigid insulators and insulator chains. They are made up of skirts, the number of which determines the voltage level.

The choice of l'isolators depends on :

1. Operating voltage ;
2. Mechanical forces ;
3. Ground pollution ;
4. Cofit.

b. **The clamp**: this is used to attach the conductor to the insulators.

c. **The spark gap**: this is used to protect the line and insulators against atmospheric surges and faulty operation, by conducting the surge to earth.

d. Lightning arrester: protects the line or the installation as a whole against atmospheric or manoeuvring overvoltages.

NB: Overhead lines are subject to bad weather, which means they need to be protected against the effects of electrical phenomena in the atmosphere. What's more, they are annoying and unsightly in large conurbations.

II.2.5.4 Underground power lines[29]

In the city of Kinshasa, electricity is mainly transmitted or distributed underground, i.e. cables are buried.

Underground lines are lines used in urban areas for aesthetic reasons, and are used for all voltages.

They are also expensive.

1. **Laying cables (housing)**[30]

In a LV distribution network, cables are laid or housed as follows:

a. **Sliced installation**

This is the simplest method, used for cables with a metal frame. In most cases, the trench is 60 cm to 1.2 m deep. In LV 0.6m under areas not accessible to cars and 1m under areas accessible to cars. This descent is made with a radius of curvature of

- 9 times the diameter for 1-conductor cable
- 8 times diameter for 2 to 5 conductor cable
- 5 times the diameter for cables with more than 5 conductors

Figure 16: Sliced installation

[29]LUZOLO.E, op.cit . p 60 - 65
[30]Ditto
[3]Ditto

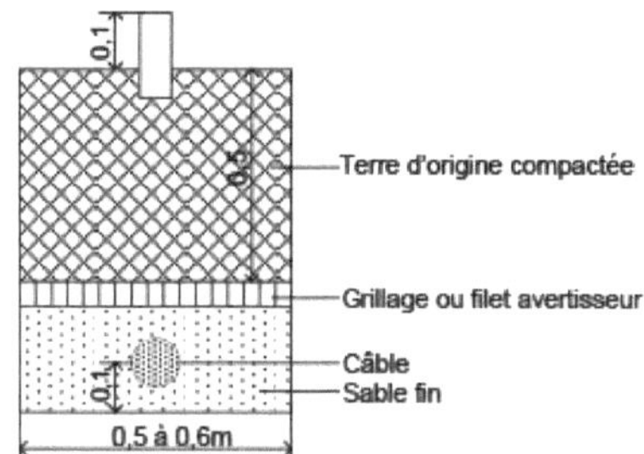

Earth of compact origin
Warning mesh or netting
Cable
Fine sand
0.5 to 0.6m

NB: At each change of direction or each crossing. Signposts must be placed.
The width of the slice will obviously depend on the number of cables.
to be installed, bearing in mind that a distance of at least 20 cm must be maintained between the various connections.
When crossing roads or railways, MV, HV and VHV cables are used.
will pass through PVC sheaths (base) embedded in concrete.

II.4.6 Bt switchgear (PS)[29][31]

1. Definition

Switching stations are network equipment
This is an electrical switchboard that collects the voltage from one or two low-voltage feeders or the general low-voltage switchboard and distributes it according to the avenues to be served.
Isolation substations are used to manage the
manual or remote-controlled.

2. Types of switchgear

There are 3 types of switchgear, including :

- ❖ The fibreglass isolating station ;
- ❖ The metal isolating station ;
- ❖ The steel sectioning station

3. Constitution

An isolating station is made up of :

- Low voltage busbars ;
- Fuse-holder or regulator bases;
- Insulators to separate the busbars from the metallic earth;
- An earth network.

[31] SNEL cvbandal

CHAPTER IV. ENERGY MANAGEMENT

SECTION 1. ENVIRONMENTAL STUDY

The environment is a set of natural conditions likely to affect living organisms. When we talk about the environment, we cannot separate ourselves from its fundamental concepts, which are made up of three important elements. These are the physical, biophysical and human elements[32] .

The first is made up of water, air and soil, the second is formed by the flora that is the wealth of Kinzono and the third is the most important element of the environment, because it is man who is also destroyed by his own culture and customs.

The environment is also considered to be natural capital, and the economy depends above all on the environment, whose function is to contribute to production (land, water) and act as a regulator.

Our environmental study concerns Kinzono, a rural locality in the commune of MALUKU which does not have the factors required for development or which is still in the process of development, and is therefore an obstacle to the development of this sector.

So if we really want progress to reach the rural environment, we need everyone to put their best foot forward and become aware of the situation in the rural environment or encourage private, government and international initiatives (investment).

Poverty is one of the criteria for under-development in rural areas, and reducing it is one of the Millennium Development Goals.

If we really want the rural environment to develop, we have to get the farmers out of their situation, i.e. install the electricity that can produce wealth, so rural electrification is vital to the country's development.

IV.2 PRESENTATION OF THE SITE

The MALUKU commune is the largest commune in the city of Kinshasa, containing at least 79% of the provincial territory.

Figure 17: Presentation of the site

[32] REC/India, Rural power distribution and energy management in developing rural countries, March 2008.

IV.2.1. Background

Etymologically, Maluku comes from the TEKE word "MALU", which means "difficulty". This is because Maluku centre was an enclave and the only way to get to the centre of Kinshasa.

The first mayor of the Maluku commune, Mr MULELE NKIE MBAMA, moved the commune's headquarters to MENKAO.

MALUKU: is an urban-rural commune in the city-province of Kinshasa with a surface area of 7.9480 km^2 .

Population of Maluku

- **POPULATION 659,953**
- The current mayor of MALUKU is: Mr PAPY-EPIANA

IV.2.2 Location

To the east: through the Kwamouth and Kenge territories and the Kwango river upstream to its confluence with the Nkole river.

To the west: through the commune of the N'sele source of the Fu-Kiene river, a straight line joining it to the grid reference marked 52/10. Followed by a straight north-south line to its confluence with the LUO river, and downstream, from the LUO river to the confluence of the N'sele and BWA rivers, which separates in particular the commune of Maluku from that of N'sele.

To the north: by the majestic Congo river, from its bridge near the mouth of the NKAO river into the Congo river with the Maindombe river (black rivers).

In the South: through the territories of KIMVULA and Kasangulu in the province of Central Kongo: starting from the confluence of the NKOLE river, a straight line joining the confluence of the Mbete and Lumiere rivers.

- KINZONO is a locality situated in the commune of Maluku and its local chief is called: MBAMA-MBUMU-JULES and his deputy.

Kinzono: located a few kilometres from Maluku

MALUKU has three main businesses

- Agriculture

- Farming and fishing

MALUKU has 10 customary groups and 7 sub-groups.

IV .3 ENVIRONMENTAL ANALYSIS

Assessing the environmental impact of energy production and consumption, together with that of related facilities.

V V.3.1 Assessment technique

We carry out a balance sheet of the current state (before the investment) and the consumption and operating balance sheet.

The gas produced by the facilities is compared with its CO_2 equivalent.

Example: 1 kg of gas produced is equivalent to 21 kg of CO_2.

The environmental impact will also be assessed in terms of the greenhouse effect.

VI .3.2 Social impact

1. Aim: To assess the effects of rural electrification on the population.
2. Some of the social benefits of rural electrification

- Reduction in rural exodus ;
- Improving treatment conditions in hospital ;
- Access to light, culture and entertainment;
- Improvements to physically demanding jobs (housekeepers);
- reducing household poverty ;
- Increase in the population of 1 household money ;
- Improving reproduction (cooking + educating children) ;
- Stimulating economic activity ;
- Use of electrical machines and equipment.

The commune of MALUKU is the largest commune in the city of Kinshasa in the Democratic Republic of Congo (DRC). The commune of MALUKU contains at least 79% of the city of Kinshasa and the other 23 communes share 21% of the territory.

MALUKU is an urban-rural commune with two vocations: agriculture and fishing. It has a total surface area of 7,984 km^2 , a population of 656,872 and is bordered to the east by the new province of Kwango, to the west by the commune of N'SELE, to the north by the Congo River and the province of Kwilu, and to the south by the territory of Kasangulu and Madimba in the province of Kongo Central.

The commune of MALUKU is the largest commune in the city of Kinshasa and is subdivided into 45 districts, including several villages embodied by the customary power of 10 groups. The commune of MALUKU is sparsely populated, with only 23 inhabitants/km^2 .

This is the commune where the natives of the Teke region are most to be found.

IV.3.3 Geographical representation and list of districts in the municipality of

Maluku

A. Geographical representation

Figure 18. Map of the city of Kinshasa

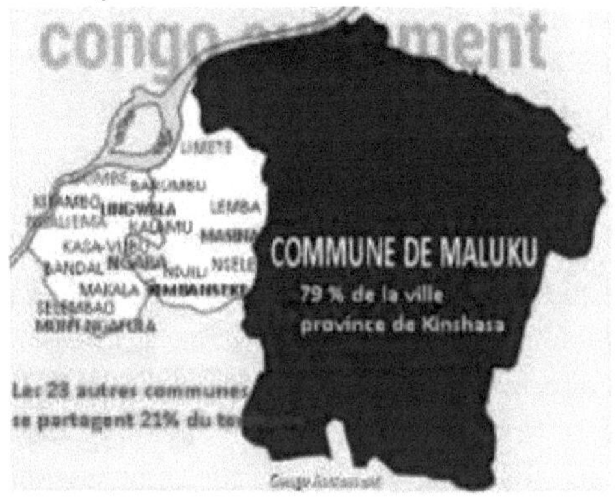

B. Table 9 The neighbourhood list

N°	Neighbourhood	N°	Neighbourhood
01	BANGO MBAMU	24	MBETE
02	BITA	25	MBOKA POLO
03	BU	26	MENKAO
04	DOKOLO	27	MOKAMO
05	DUMI	28	MOLOKAYI
06	IMBIA	29	MONACO
07	INGA	30	MONGATA
08	INKIENE	31	MOSABU
09	KIKIMI	32	MPO-KINSELE
10	KIMPETI	33	KINTA KISANKULU
11	KIMPOKO	34	SHEEP
12	KINGANKATI	35	NDAKO PEMBE
13	KINGAWA	36	NGAMANZO
14	KINGUNU	37	NGANA
15	KINTA	38	NGUMA
16	KINZONO	39	NKOMO
17	MAES	40	NSUNI
18	MAI-NDOMBE	41	NZAMU
19	MALUKU	42	SUALEPU
20	MAMBUTU NKA	43	WASSA

21	MANDUNGU	44	YOSSO
22	MANGENGENGE	45	YUO
23	MBANKANA		

As the commune of MALUKU is large and has several districts, we are not going to talk about all the districts, but we are going to study the locality of KINZONO, which is a new district in the commune.

C. REALITY OF THE RURAL ENVIRONMENT OF THE KINZONO DISTRICT

1. History

The locality KINZONO is a new district of the commune of MALUKU, at the time it is only one district which was named INKIENNE which is vast, before decoupage, it was composed of 11 localities one of district of MALUKU the largest which was directed by BOLINGO MONDEMBETO, towards the years 2017, it was the very first decoupage in the commune MALUKU which gave birth to 6 districts and follows to the hear of the commune the governor and the burgomaster find good to decouper again the commune, towards the years 2020 was the second decoupage of which one has birth of KINZONO also which counts 14599 inhabitants.

2. Population

The rural population is the total number of people living in rural areas. The rural environment is defined as 1 set of Fokontany whose proportion of the population exergant.

In KINZONO, on the other hand, their way of life is based more on agricultural activities (farming, livestock rearing and fishing), which accounts for over 70% of their income.

Agricultural activity is the set of individuals formed by agricultural households. This is a household in which one or more active members carry out one or more agricultural activities on a principal or secondary basis.

Some of them do trade, but their businesses are based more on agricultural products.

3. Economic situation

The different activities in this area are :

- Agriculture ;
- Breeding ;
- Fishing;
- Forestry

But what interests us most is agriculture and fishing.

a. Agriculture

The agricultural population is defined as all individuals living in rural areas and belonging to an agricultural holding. It is estimated at 70%, 35% of whom are women and 25% men.

Most of them are parents. There are almost as many 20-year-olds doing this activity too, because it's the most popular in the neighbourhood.

The agricultural population aged 12 and over is considered as the potential active population in agriculture because from this age, the individual can contribute to agricultural activities. This age group represents 70% of the agricultural population and has more women than men, with a ratio of 95 men to 100 women. At this age 77% are employed, 5% are unemployed and 18% are inactive. The inactive include the elderly and disabled, housewives, students and schoolchildren, and children who have not taken part in farming activities. The agricultural population thus has an index of economic independence. Almost 70% of the rural population of KINZONO are farmers.

b. Fishing

Fishing also plays a key role in Kinzono, and is practised by the majority of the population (men). Fishing has an impact in Kinzono because of Lake Kwango, which separates it from the Bagata territory in Kwango province. Thanks to this lake, the people of Kinzono also fish[31] .[33]

c. Water quality

Water quality (taste, smell and colour taken together) was appreciated by 63.9% of households. The colour of the water was the main reason for this level of appreciation. In addition, 85% of respondents said that the water was suitable without boiling. But this water comes from springs and wells.

d. Soil quality

The quality of the soil is perfect, it's also their great wealth, thanks to their soils, they do business as usual.

IV.3 OBJECTIVE AND SOME DEFINITIONS OF THE RURAL ENVIRONMENT

IV.3.1 Study objectives

- Identify the different sources used to supply rural and urban-rural areas;
- Determine the means of transmitting this energy to the consumer;
- Evaluate the billing method for rural consumption.

IV.3.2 Definitions

a. Electrify

This is the supply of electrical energy through the installation of an electrical source, line or network.

[33] Op.cit. pp. 20 - 30

b. Rural

Relating to the countryside, as opposed to the city.

c. Rural Electrification

This is the set of techniques used to supply the countryside, taking into account its specific features and those of its inhabitants[34] .

IV.3.3 Characteristics of the countryside or rural area

- Rural areas are less populated or have low population density and low purchasing power;
- The population is mainly active in agriculture;
- Trade, industry and administration reduced to a minimum;

IV.3.4 Rural electrification difficulties

Rural electrification is characterised by :

- A dispersed load connected to sources or transformers by transmission sources;
- A very high investment cost for a sparsely populated, low-income population.

Socio-economic factors pose a complex problem in determining the real cost of billing.

In addition, very high energy costs will have an impact on the cost of agricultural products exported to urban centres.

IV.3.5 STATUS OF THE RDC

Despite all its energy potential, the DRC has very little electricity.

Only 6,200,312 of the estimated 9,421,569 households have electricity, i.e. 6.48%.

The provinces of BANDUNDU, KASAI ORIENTALE and MANIEMA have a very low electrification rate of around 1%, while KONGO CENTRAL has the highest rate at 11%.

Table 10: Coverage rate

ITEM	PROVINCE	COVERAGE RATE
1	BANDUNDU	0,12%
2	KONGO CENTRAL	11%
3	ECUADOR	0,68%
4	EASTERN PROVINCE	2,69%
5	WESTERN KASAI	0,45%

[34] REC/Mode, Rural power distribution and energy management in developing, March 2008.
- Perc, rural and cross-border electrification, Kinshasa, 2010
- KHUZAMA and VIKOIS, prepayment master 2010.
- KHUZAMA and VIKOIS, si single-phase High-distribution system 2010.

6	EASTERN KASAI	0,14%
7	KATANGA	4,43%
8	KINSHASA	40,67%
9	MANIEMA	0,1%
10	NORTH KIVU	1,47%
11	SOUTH KIVU	4,43%

STRENGTHS AND WEAKNESSES OF THE DRC :

- Very high hydroelectric potential, 4eme position in the world with 600 billion kWh;
- Diversity of energy sources (solar, hydro, biomass) ;
- 80.9% of men and 54.1% of women are highly literate;
- Very active informal sector;
- The population is constantly growing, with 6 people per household and a growth rate of 2.8%.
- High rural exodus (urban population reaches 44%)
- Very indebted poor countries
- The population's purchasing power is very low;
- Recurring conflicts and wars.

Figure 19: Structure of a rural electrification network

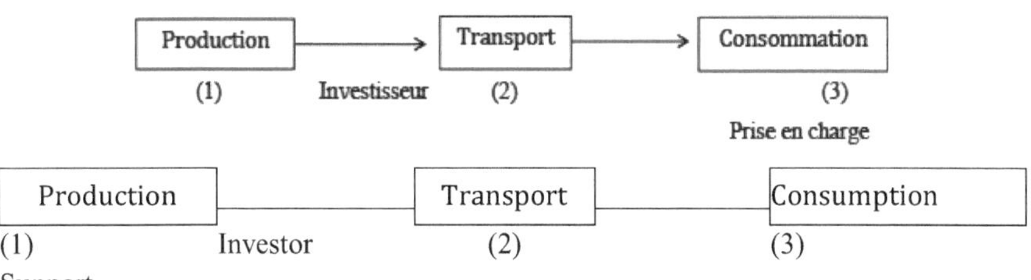

IV.4 Hydraulic micro power station

❖ **Its role and investment**

Role of micro-hydro power station

- Produce
- Satisfy
- Creating value
- Promoting development 1. Is the need urgent (long term, medium term)?

2. Are funding sources available?

There has to be a permanent watercourse where a micro-power station can be installed.

Turbines of the BANKI type seem to be better suited to rural areas because of their simplicity of manufacture.

Figure 20: Hydraulic micro-power station

Rural and urban-rural areas require facilities that are expensive per head of dwelling, but the population is generally on a modest income and needs to be managed in such a way as to take responsibility.

The various stakeholders: the population, the State and investors must be involved in covering these different costs.

The stakeholders involved in or driving this implementation.

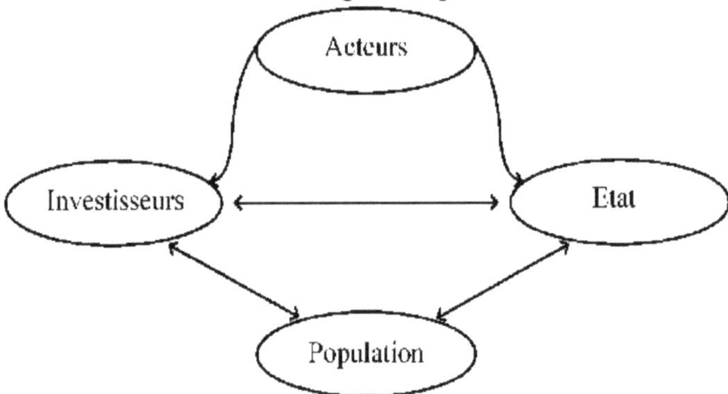

- Village 700 inhabitants
- Micro power plant that provides families with distribution capacity and impact
- One house = power = 4 kwh the average power distributed per house 4kwh.

Knowing that the total capacity is 280 to 300KVA.

IV.5. IMPORTANCE OF ELECTRICAL ENERGY

IV.5.1 INTRODUCTION

In just a few decades, mankind has burnt up a large part of the energy reserves

that the planet has built up over millions of years. We are draining the earth of its resources, the atmosphere is becoming enriched with CO_2, the greenhouse effect is increasing, ecosystems are deteriorating, and the climate change that was predicted is now a reality.
Today, electricity is useful for every aspect of our daily lives. Electricity helps us in the home, for care, for consumption.
The current state of the energy sector in the Democratic Republic of Congo, and in particular the locality of KINZONO in the commune of MALUKU, shows a very low level of energy supply to the population. The slackening of the implementation of adjustment policies in 1990 did not allow for the development of the energy sector as a whole, mainly because of the deterioration of the main economic parameters that followed. More than 10 years on, the situation is alarming.
The inadequacy of private initiatives in the energy sector and the lack of funding are some of the obstacles to the recovery of the energy section in the Democratic Republic of Congo. For more than 40 years after the country's independence, the state of the sector shows a very low rate of energy supply to the population, following the absence of appropriate programmes to improve the dysfunctions in the daily lives of the Congolese people, particularly those living in the KINZONO locality in the commune of Maluku. Only 10% of the population have access to electricity, which poses a serious problem.
Energy is a difficult concept to learn. It is stored in matter and only manifests itself when it is transformed. It is at this point that we perceive its effects: the heat supplied by a lamp, energy never disappears; it is transformed from one form into one or more others.
As energy is the driving force behind all socio-economic development, it is well known that energy services make a significant contribution to driving economic and social development, as well as being vital to all human activities in order to achieve better living conditions.
Electricity is the form of energy that powers most of our daily activities, and is one of the vital elements in a country's development.
It's hard to imagine today's world without electricity; the applications of electricity are ever more numerous and accompany new inventions and technological advances; as a result, electricity consumption increases every year. This leads us to say that if the energy sector is not well managed, it will be difficult for development.

IV.5.2 Energy types

These include renewable forms such as :

- Solar energy ;

- Wind energy ;
- Hydropower ;
- Wave energy ;
- Tidal energy ;
- Marine energy ;
- Geothermal energy.

IV.5.3. HOW ELECTRICAL ENERGY WORKS

Electric energy is any energy that is transferred or stored using electricity.

Electricity is not a primary energy source: it requires another energy source in order to be produced. Electricity is often considered to be a clean source of energy, as it emits no carbon dioxide or fine particles. However, depending on the primary energy used to produce it, it can have a significant environmental impact:

Thermal power stations and nuclear reactors, for example, have a significant environmental impact. Renewable sources of electricity such as wind power, solar energy, hydroelectricity, biomass and geothermal energy are reputed to be clean, but their use must also be subject to strict conditions to ensure an effective energy transition.

IV.6. ENERGY PRODUCTION

Electricity is energy generated by the movement of positively or negatively charged particles. Electric current is direct when the electrons move in the same direction in the conductor, and alternating when they move in one direction and then in the other at regular intervals called cycles.

Electricity is now used all over the world. Its main uses are heating, hot water production, lighting, and powering a growing number of household appliances and electronic devices such as computers, television, etc.

IV.7. CHALLENGES AND BENEFITS OF ELECTRIC POWER

Electricity is now essential to the functioning of the global economy. Electric energy has many advantages, but it also has its limitations, which means that the various players in the sector will have to face up to the challenges of the coming decades.

Electricity is a relatively inexpensive energy to produce. During its transport and consumption phases, it can be considered a fairly clean energy, as it has a favourable carbon balance and emits very few particles.

IV.8. ELECTRICITY ISSUES

The issues surrounding electricity mainly concern its storage. Electricity is difficult to store in sufficient quantities and at a reasonable cost to meet the energy needs of a population. Research is also underway into the use of hydrogen as an electricity storage medium.

CHAPTER V. CONCEPT OF PROFITABILITY

V.1 GENERAL INFORMATION ON PROFITABILITY

The concept of profitability is closely linked to that of profit and applies in particular to businesses and to any other investment. In general, the aim of any business is to be profitable[§§§§§§§§§] .

For the shareholder, the result is first and foremost the dividend that is paid to him for the financier. For the State, it is profit before tax. From the point of view of performance, it is the value added by the company.

V.1.1. Definition of the concept of profitability

In corporate financial linguistics, the word profitability is used in several senses:

- It is often used as a synonym for profit in absolute terms: "the company did not make a profit in this financial year" almost means that the company did not make a profit in this financial year. This is a completely erroneous concept. Profitability is not an absolute value. It is a relative concept, the quotient of a ratio ;
- Strictly speaking, profitability has two specific characteristics: it is a capacity, a potential return. It is therefore the measure of the remuneration of the providers of capital, the owners of the business. This remuneration corresponds to what is left over, i.e. what was paid or recorded when it was paid, or which corresponded to an idea of substance: in particular, depreciation and provisions.
- In the broadest sense, it refers to the ability of all kinds of capital to bring in money. There are several types of profitability, including :

> Economic profitability, defined as the ratio between profit and the capital employed to obtain it.

> Commercial profitability, which is the ratio of profit or loss to sales.

Generally speaking, the term "profitability" refers to the ability of capital to generate a balance which, depending on the circumstances, is called profit, profit, surplus or "fruits". Whatever the economic system, this notion of profitability appears to be the criterion of economic efficiency.

According to MBANGALA MAPAPA (2009), assessing the concept of profitability has multiple facets, each of which relates to a specific objective in the contribution of value created. The latter is defined by the extent to which the company achieves its objectives. Financial analysis focuses its analysis on one of these objectives. This is why it is important to have well-defined objectives and the resources deployed to achieve them. Generally speaking, a company's main objective is financial profitability[*********] .

A company's profitability is measured as the ratio between the results obtained

---

[§§§§§§§§§] SAMBA V, Comptabilite analytique OHADA, 2ème edition, Paris 2020, pp 10 - 20

[*********] MBANGALA MAPAPA, Comptabilite generale OHADA, Paris, 2nd Edition, 2010.

by the company and the resources used to achieve these results.

V.2. THE ROLE OF PROFITABILITY

It is a key element in assessing company performance.

V.3. IMPORTANCE OF PROFITABILITY

There are a number of reasons why it is important to spend time analysing a company's profitability and whether it is doing well:

- To ensure that your business lasts for a long time, if your profitability is good, you'll have enough money to keep your business running smoothly.
- So that you can invest if the profitability is such that you manage to make a profit, you will then be able to reinvest this surplus money in your business.
- So that you can improve the growth of your business naturally, if you manage to invest in your business thanks to the profit generated by the good profitability of your business, this will generate growth.

V. 3.1 The importance of profitability

It refers to the ratio between the result obtained and the capital invested (or committed). From a purely financial point of view, the objective of a financial or operational investment is to maximise profitability.

V.3.2. Types of profitability

1. Economic profitability

Profitability is the company's ability to transform this capital into profit, whatever its source.

2. Financial profitability

Measures the return on equity capacity only.

V.4. TABLE 11: PARALLELISM BETWEEN ECONOMIC PROFITABILITY AND FINANCIAL PROFITABILITY

Economic profitability is primarily of interest to managers	Financial profitability primarily of interest to shareholders
Ratio between operating profit and the resources invested in the company's activities to achieve it.	The ratio of a company's profit to its shareholders' equity
Used by managers and lenders to assess and compare business performance.	Company directors have an interest in improving the company's financial profitability because they are elected by the shareholders.
To take account only of the company's normal activities	The higher the rate of return on equity, the more
Operating profit is used, excluding financial and exceptional items. Capital employed corresponds to the value of	The company will find it easier to raise funds on the markets.

gross + the value of operating working capital.	
Economic profitability _Re s u l tat d ie xp l oitati o n c a pita u x i n ee s ti s Independent of the company's financing structure because the	Financial profitability _Re s u l tat d e l i e xe r ci ce c a pita u x p ro p r e s s Takes into account the financing of the company, as financial charges
operating profit is independent of the way in which the company is financed.	are included in the result (they reduce it)

A business project is profitable if it generates regular and substantial income for the entrepreneur, and if it ensures the payment of a substantial capital sum when the business is sold (capital gain). In the case of business start-ups, projected profitability is generally measured by the profits generated and the income paid to the entrepreneur.

To calculate the profitability of a start-up project, it is important to draw up what we call a business plan.

You can only measure the profitability of your business start-up project if you have a document that provides a basis for calculating indicators. You need to draw up a business plan, or at least a financial forecast.

This represents the financial part of the business plan. It consists of a series of tables (income statement, financing plan, balance sheet, cash flow budget) which highlight key figures such as operating profit.

For your calculations to be reliable and usable, you need to ensure that your financial document is complete. This involves a number of things. First of all, you need to estimate your turnover as objectively as possible. To do this, base your estimate on the results of your market research. Don't overestimate it. Next, list all the costs you are going to encounter in your project. Assess them accurately.

Don't forget to include one important point in your business plan: your remuneration.

V.5. MEASURING THE ACHIEVEMENT OF PROFITABILITY

Your project will generate expenses, some of which will depend on the level of business (these are known as variable costs), while others will have nothing to do with it (these are known as fixed costs). The relationship between fixed costs, variable costs and turnover is very simple:

- There are no variable costs in sales. But you will have fixed charges to pay;
- With a turnover, you face not only costs but also variable costs.

Fixed costs: property rents, salary, social security contributions, which do not depend on the volume of business.

Variable costs: purchase of raw materials and supplies, transport costs on purchases and sales. These depend on the volume of activity, REGIDESO invoice, SNEL invoice.

V.6. VERIFIABLE OBJECTIVE INDICATORS FOR OUR PROJECT

An idea often turns into a project. But a project only has a chance of succeeding if it is variable. The viability of a project is fairly complex to measure. In reality, it is assessed in several areas: the technical area, the strategic area (strategic and operational management), the economic area, the financial area and the psychological area.

Our project is viable because we have examined two very important parameters: firstly, the relevance of our offer, which must provide a solution to an existing problem, because without a problem, there is no demand and therefore no offer to make; secondly, you need to validate the product/target pairing, i.e. make sure that the offer (and its added value) matches the customer base.

Our consumers are psychologically ready to buy our product, but they are also aware that the environment has shown us this great opportunity.

In most cases, market research is used to verify all of these questionnaires on a carefully selected sample of potential customers, which has enabled us to gather information on this subject.

We have set up what are known in the jargon as communication channels within our organisation - a communication plan: advertising, word of mouth, press articles, etc. - because there's an adage that goes: "a good idea won't work if the public doesn't know about it". Our distribution channel is really secure and, above all, we are in a non-competitive market (blue ocean).

In setting ambitious targets, we used the SMART method (specific, measurable, achievable, realistic and time-bound).

V.7. PROFITABILITY FACTORS

Several factors influence profitability:

V.7.1 External factors

- Tax burden (ratio of tax revenue to income).
- Currency devaluation
- The economic situation
- State intervention
- Falling income
- The company's social climate
- Remuneration policy
- Motivation
- The company's internal growth

V .7.2. Internal factors

Profitability is a key factor in assessing company performance. Profitability plays an important role in the life of a company; it ensures the company's survival and enables it to show its financial independence.

V .8. CALCULATION OF PROFITABILITY

V .8.1 Further calculation of the profitability of a project

Profitability is generally measured by comparing performance (earnings management) with the resources invested to achieve it.

Table 12: representing the complexity of profitability calculations.

Indicator	Calculation formula	Utility
Return on equity	Net profit/equity	Measuring the return on each unit of money invested in the company's capital
Operating margin	Operating result (sales)	Determine the percentage of sales that turn into wealth for the company
Net rate of return	Net profit/sales	Expresses overall profitability
Earnings per share	Net profit/ number of shares	Corresponds to the dividend potentially attributable to each share

In the preliminary project phase, the economic feasibility study consists of calculating profitability on a macroscopic basis. However, this profitability calculation stage is necessary to assess the risk and take the decision to launch the project, by also comparing the forecast calculations with the reality on the ground.

V .8.2. The different results measuring profitability

The income statement provides a quick overview of the profitability of operating activities.

V .8.3. Operating profit

Resulting from the difference between operating income, essentially sales and certain subsidies, and operating expenses (overheads, personnel costs, etc.), a measure of the performance of the day-to-day running of the business, excluding financing policy.

V .8.4. Net profit

Determines the company's overall annual performance, for all cycles combined: operating, financial and exceptional. This is obtained simply by adding operating, financial and exceptional expenses to income of all kinds.

Similarly, the resources deployed by a company can be measured by :

Balance sheet total: total assets measure all the goods and rights used by the company to produce.

Equity capital measures all the stable financial resources immobilised by the company for production purposes. The company's share capital, which measures all the financial resources advanced by the company's shareholders[35] .

A. Commercial or business profitability

This ratio expresses a company's profitability as a function of its volume of business. It is calculated as follows:

$$\text{Commercial profitability} = \frac{\text{Résultat net x 100}}{\text{Chiffre d'affaires}}$$

We then determine the company's margin rate, which enables us to estimate the company's future profit as a function of the variation in its volume.

This profitability is obtained by comparing the profit with the pre-tax turnover, which is representative of the business. The most widely used business ratio is the "net margin", which indicates the proportion of the company's business accounted for by profits.

$$\text{Net margin ratio} = \frac{\text{Résultat net}}{\text{CAHT}}$$

B. Financial profitability

This ratio measures profitability in relation to the capital invested in the company and can be assessed in two ways:

- By the rate of return on equity
- By the rate of return on stable resources

* The rate of return on equity. It is determined as follows:

$$\text{Financial profitability} = \frac{\text{Résultat net x 100}}{\text{Capitaux propres}}$$

It is used to measure the profit earned on the funds provided by shareholders. Rate of return on equity

$$\text{Commercial profitability} = \frac{\text{Résultat net x 100}}{\text{Chiffre d'affaires}}$$

$$\frac{\text{Bénéfice distribuer}}{\text{capital}}$$

Or

* Rate of return on stable capital or sustainable resources.

$$\text{Return on permanent capital} = \frac{\text{Bénéfices+intérêts}}{\text{Ressources durables}}$$

C. Economic profitability

This indicator measures profitability in relation to the fixed assets used by the company to produce. It is calculated as follows:

Economic profitability = $\frac{\text{Résultat net x 100}}{\text{Investissement total}}$

This ratio is a more relevant indicator of profitability measured in terms of the efficiency of the production process, also known as the "profitability of ....". It expresses the profitability of all the capital employed. It can be assessed in several ways:

Commercial profitability $=\frac{\text{Bénéfice+intérêts}}{\text{Total actif}}$

Return on capital employed $=\frac{\text{Bénéfice net}}{\text{Total actif}}$

D. Overall profitability

It measures the profitability of all the assets used by the company. It is calculated as follows:

Overall profitability $=\frac{\text{Résultat net x 100}}{\text{actif total}}$

V .8.5. Profitability assessment

Estimating a company's profitability is therefore a good indicator of its efficiency or performance in its production function. For this reason, calculating a company's profitability should be accompanied by a comparison of its level of profitability with that of its main competitors, or with the profitability it has achieved in the past.

V .8.6 PROFITABILITY THRESHOLD

V .8.6.1. Definition

The break-even point for a project or a company is the level of net sales for which there is neither a loss nor a profit, and which will generate a profit once the current operating costs for the period have been included. Conversely, if it does not reach this level, it will make a loss, whatever the nature of your business (industrial, commercial, craft, self-employed or agricultural).

The break-even point is the volume of net sales for which the margin on variable costs exactly covers the fixed costs, the break-even point is synonymous with critical sales; break-even point, neutral point, the following follows:

- The result is nil when the turnover is equal or equal to the profitability threshold.
- The result is positive (profit) if sales exceed the break-even point.

V.8.5.2 Objectives

For a project, for a company, the break-even point is a "flashing light" that is observed at all times. So, in a difficult moment, managers may have to ask themselves whether to continue or stop the business.

It allows you to :

- Identify the level of activity below which you should not go, or the level at which your business becomes profitable;

- Measuring operational risk ;
- Anticipate the results to be achieved in the future;
- Understanding the impact of changes in costs on profitability ;
- Know how much of a drop in activity the company can withstand with the safety margin it has.
- The partners you can approach will study your profitability threshold: banker, investor, partner, etc. For the banker, the break-even point is an indicator of the risk involved in your project. The lower the break-even point, the greater the risk.
- For investors, it's a matter of studying when your company will stop losing money, and indicating how long it will be before their investment pays off.

V.9. CALCULATION OF PROFITABILITY THRESHOLD

V.9.1. First method of calculation

For the project to reach break-even point, the net operating result will have to be zero.

a) Margin on variable costs = fixed costs
b) Margin on variable costs-fixed costs = 0
c) Net sales -(variable costs + fixed costs) = 0

Example

We have the following elements at our disposal:

- Net sales	12.000.000$
- Variable cost	- 8.000.000$
- Margin on variable costs	4.000.000$
- Fixed cost	- 3.500.000$
- Net operating profit	500.000$

With fixed costs of $4,000,000, the break-even point is reached with a turnover of $12,000,000.

First of all, we need to distinguish between fixed and variable costs:

- Firstly, fixed costs (these are costs whose level does not vary with the level of activity and which your company will have to bear whatever its turnover.
- And secondly, variable costs (those whose amount varies in proportion to fluctuations in business activity).

Fixed costs are generally considered to be :

- Rents on movable and immovable property ;
- Insurance premiums (professional indemnity, guarantee)
- Postal charges (stamps) and telecommunications charges (telephone, internet connection charges)

- Fees paid to external service providers (chartered accountant, lawyer, notary, tax advisor)

Conversely, the main variable costs are :

- Purchase of goods
- Subcontracting costs
- Production costs of products sold
- With fixed costs of $1, the break-even point is reached with a turnover of $\frac{12.000.000}{4.000.000}$
- Since fixed costs are worth $3,500,000, the break-even point is reached with a turnover of $\frac{12.000.000 \times 3500000}{4.000.000}$ break-even point = $10,500,000

The break-even point can be determined using the following formula:

$$\text{Break-even point} = \frac{\text{chiffre d'affaires net} \times \text{coûts fixes}}{\text{marge sur coûts variable}}$$

- Second method of calculation

Designed by :

SR: break-even point

Sales: net sales

CF: fixed costs

CV: variable costs

MCV: margin on variable costs

TMCV: margin rate on variable costs

Profitability threshold $\frac{\text{coût fixe}}{\text{Taux marge sur coûts variable}}$

If expressed as a quantity

$$\text{Profitability threshold in terms of quantity} = \frac{\text{coût fixe}}{\text{Taux marge sur coûts variable}}$$

Or

$$\frac{\text{coût fixe}}{\text{MCV unitaire}}$$

T

$$\text{Dead centre} = \frac{\text{Seuil de rentabilité} \times 365 \text{ jours}}{\text{chiffre d'affaire net}}$$

This formula is applied when the project has been cancelled, i.e. 12 months.

V.9.2. Improvement of the profitability threshold

Three variables can be used to improve the breakeven point[36] :

- **Use variable costs rather than fixed costs;**

This strategy consists of making the most of your costs, i.e. ensuring that your business incurs as few fixed costs as possible in favour of variable costs. In this way, your cost structure will follow the trend of your business.

However, we have to be careful: this technique has its limits, as it will lead you to subcontract as many operations as possible and should not, under any circumstances, lead you to outsource your company's key skills.

- **Reduce the weight of fixed costs**

This method consists of controlling the amount of fixed charges by deferring certain expenses, for example, or by identifying less expensive substitutes). business at home (then find premises when your company's financial resources allow you to do so) or find premises in a slightly less sought-after area to save money.

- **Improve your margin on variable costs**

You can also act on the "traditional" levers of profitability: your turnover and your margin on variable costs. It might be possible to increase your selling price without affecting your order volume? Or increase your customers' average basket? As far as variable costs are concerned, try to negotiate discounts and rebates with your suppliers by committing yourself to an order volume.

CHAPTER VI. ANALYSIS OF RESULTS

Analysing the results gives the manager a quantitative idea of progress or management expectations.

VI.1. DIAGNOSIS[39]

- Diagnosis is the reasoning leading to the identification of the cause (origin) of a fault or problem based on symptoms; an evaluation process.
- Detection of this state of functioning of this state, it consists of identifying, analysing, it is important to make the diagnosis, because it gives us an idea of the result.

The main purpose of the analysis and diagnosis based on the income statement is to understand the formation of net profit.

This result can be defined as the difference between the amount of expenses incurred by the company and the various income received. A profitable result is essential for a company, because it has a favourable effect on the conditions that guarantee its survival and development.

From the point of view of financial analysis, the result is the essential element for assessing the company's profitability and judging the effectiveness of management (quality management).

To assess a company's ability to generate profits and internal resources, it is necessary to analyse the operating and financial results in greater depth, and to eliminate the influence of exceptional operations. Indeed, it is not an encouraging sign for the future of a company or a project if it finally succeeds in generating a profit thanks to a capital gain realised on the sale of a non-current business activity.

VI.2 OBJECTIVES

The objectives of the diagnosis based on the income statement are :

- The contribution of objectively verifiable synthetic indicators capable of revealing trends.
- Improving understanding of the content and size of financial packages.
- The expression of a result-oriented action.

Cash flows related to the company's operations during the period and presents the results generated.

The income statement is structured on four successive levels: Operating activities, financial activities, non-recurring activities and tax.

The structure of the profit and loss account in SYSCOHADA (Accounting System) for the Harmonisation of Business Law in Africa is broken down into the following main sections:

---

[39] CHEVALIER. A et ROLLET, J. Organisation industrielle, production, entretien et manutention DE LA GRAVE, Paris, France, 1975

TABLE 13: SIMPLIFIED INCOME STATEMENT

Expenses	Products
Operating expenses A: Operating result	Operating income
Finance costs B: Financial result	Financial income
Charges H.A.O	H.A.O. products
D: Employee profit-sharing	H.A.O. products
E: Tax/Loss	

Source : SYSCOHADA

VI.3 FEASIBILITY

In project management, a feasibility study is a study that aims to check that the project is technically feasible and economically viable. Broadly speaking, a feasibility study can be divided into the following sections: technical, commercial, economic, legal and organisational.

Running a business often involves identifying the projects best suited to the chosen strategy. The feasibility study is used to determine whether potential projects are feasible, and plays an important role in the choice of projects.

Before embarking on the implementation of a project. It is important to ask whether the project will be profitable, whether it is technically feasible or whether the company has the financial resources, skills, strategy, operational capacity, etc. to implement it. The feasibility study has several objectives:

- Measuring the objectives to be achieved,
- Evaluate the conditions necessary for the success of the project (timing, equipment, skills, funding, etc.),
- Plan the implementation of the project,
- Study the different possible scenarios.

The feasibility study is usually carried out just after the project has been defined. Carrying out a feasibility study once the project's purpose has been defined, or once the main operational choices have been made, makes it untimely.

In many cases, the project sponsor does not have the necessary resources to carry out a conventional feasibility study.

Validating the feasibility of your project is an essential step in avoiding wasting resources on a project that may not succeed.

As costs are the accumulation of charges on a product, several costs can be used to calculate the price. These are fixed costs, variable costs and average costs. By analysing these costs, you can get a better idea of the price to be set, including operating-related exchanges.

Carrying out a feasibility study requires more rigour and method (strategy).

- We need to move forward in several stages;

- Assessing project needs ;
- Evaluate the financial cost of the project;
- Study the possible scenarios;
- Choose the most suitable scenario.

VI.4 Estimated sales costs

This is a very delicate stage in the life of a product: Setting a price.

In fact, the price is not fixed by chance and must correspond to a number of criteria.

The aim is to be able to sell, to be profitable and ensure the company's long-term future.

When it comes to setting prices, analyses generally distinguish between two factors: costs and competition.

This cost will depend on the variables involved in transport and distribution to the customer.

VI.5 Competition

The prices adapted by competitors are an important element in setting prices, in this case we don't have the competitor but rather to reflect on the purchasing power of the inhabitants.

The estimated cost is $0.01 for 1Kw

$$Rendement = \frac{les\ frais\ d'abonnement\ X\ par\ le\ nombre\ d'habitants}{Charges\ d'exploitation}$$

Investment cost: $400,000 incl. VAT

The cost invested and the determination of profitability will enable us to conclude whether the project is profitable or not.

After a detailed study of our project and our investigations, we are able to increase production capacity and supply more to a large part of the Congolese population.

HNB strategic model

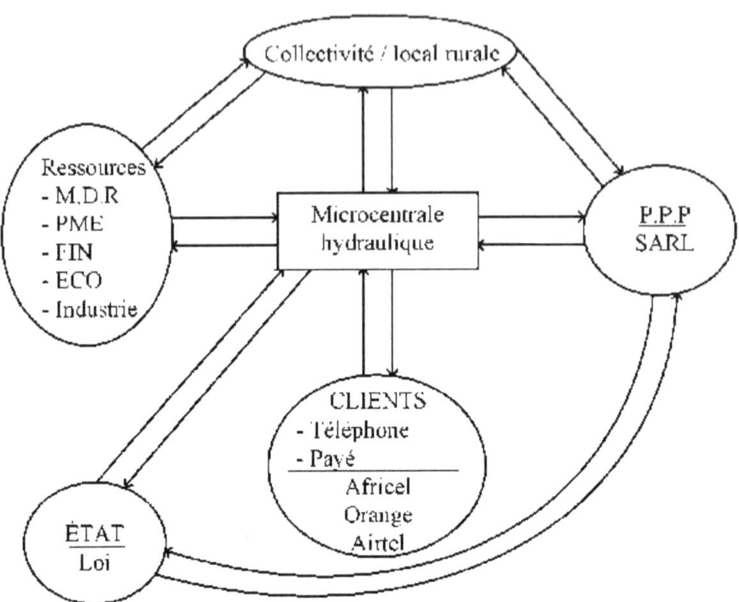

Legend:

- HNB: Hubert NTUMBA BULELA ;
- P.P.P: Public-Private Partnership
- SARL : Société par Action Responsabilité Limitee ;
- MDR: Ministry of Rural Development ;
- MPME: Ministry of Small and Medium-sized Enterprises;

VI.6.1. Notion

The project we are presenting covers several areas in which it is advisable to put in place a tool to facilitate and interpret our project both internally and externally, namely the business plan.

VI.6.2. Definition

> The business plan is a written document used to formalise a project to set up a company or develop a business.

> A business plan is a document (or a set of documents) in which a business project is presented. As its name suggests, a business plan presents an economic and strategic development plan for a company.

> It is also a summary document that enables an entrepreneur to present the ins and outs of a project simply and effectively.

> It is also a document designed to convince a decision-maker to invest in an entrepreneurial project.

This document describes how the business will generate profitability, setting out all the means envisaged to deliver revenue, to whom for what purpose and by what means.

VI.6.3. Purpose

The aim of the business plan is to provide a framework for the business, i.e. to be able to make changes to the initial forecasts in order to be as realistic as possible. However, the business plan should not be confused with a market study.

A business plan is a necessary step before setting up a company. It is intended for the company's investors and employees, and is a document that influences the decisions of the partners. Its main purpose is to give details of the project (summary, team, business idea, etc.).

1) Market figures
2) Future trends
3) Success factors

1) THE PROJECT

1) Presentation
2) Value proposition
3) The project leader

2) MARKET RESEARCH

1) Walking segments
2) Swot analysis
3) Competitor analysis
4) Study of the competition
5) Competitive advantages

3) THE STRATEGY

1) The action plan
2) The business model framework
3) Marketing strategy
4) Marketing action plan
5) Risk management
6) Why the project is viable

4) CONCLUSION OF THE FINANCIAL FORECAST

1) Provisional profit and loss account
2) Forecast balance sheet
3) Forecast cash flow budget
4) Performance indicators
5) Calculation and analysis of WCR
6) Financing plan

7) Outgoing investors

The business plan is structured in 5 parts, which flow naturally into each other to make it easier to read.

V I.7 MARKET OPPORTUNITY

Given that electrical energy contributes to development, with our survey and observation methodology 98% of the population is interested in this[40] .

The turnover for the installation of a micropower station is generally between €65,000 and €120,000, for a good distribution of energy.

In terms of profitability, our SME spends around 30% of its income on equipment and raw materials, 20% on rent and administrative costs and 5% on marketing and communication.

The profitability of a micropower station is generally between €10,000 and €25,000.

This part shows that you know your walk, its size, its value and its dynamics.

V I.8 TRENDS

Given the need for energy, demand is increasing in the environment. Analysis of these trends shows that you are aware of changing consumer habits and will develop a project that adapts to the needs and desires of the market.

V I.9 SUCCESS FACTORS

Examples of business plans

- Location and accessibility
- Quality-price ratio for consumption billing
- Importance of electrical energy
- The implementation of this micropower station will create wealth

This part of the business plan shows that you are aware of the factors that will enable you to build a sustainable and profitable business.

V I.10 THE PROJECT

After the background, here's the presentation of our project.

Here, we're presenting our project in the form of a target, so we're highlighting some of the strengths of our future business, which involves setting up a micro-hydro power station to combat poverty.

V I.11 THE PROJECT OWNER

The project owner is originally from the DRC province of Kinshasa, I studied commercial and financial sciences, specialising in strategic and operational management at CEPROMAD University where I learnt all the skills relating to strategic and operational management.

So I've got the managerial skills needed for this new project, and as a student at CEPROMAD university's doctoral school, I'm sociable and ambitious,

[40] Op.cit pp 100 - 1010

motivated and go all the way with my projects.

I've been a project leader for four years now, working as a 2[ieme] assistant in an official local institution, taking on a number of academic roles with an entrepreneurial culture, and capable of leading a project.

V I.12 ON-SITE STUDY

V I.12.1 MARKET SEGMENTS

Our project is located 120 km from the city centre in the urban-rural area known as KINZONO, which is characterised by a lifestyle that is the opposite of that in the city. It is the households and income-generating activities that are of most interest.

VI.12.2 SWOT MATRIX

Here, we're going to ask ourselves a few possible questions:

1. What are the strengths and opportunities that will enable us to gain market share?
2. What factors could threaten our development?

VI.12.3 FORCES

- No electricity in the KINZONO district, not even in the surrounding area
- Demand supported by market research
- Experience of the project leader
- Support from local residents
- No electrification of this urban-rural area

VI.12.4 WEAKNESS

- Partly dependent growth
- Lack of notoriety
- Weak financial capacity at start-up
- Substantial initial investment

VI.12.5 OPPORTINUTES

- Partnerships with economic operators
- Events with local producers
- Advertising on social networks
- Creating a long-term franchise
- Excessive development of

The SWOT matrix is the preferred structure for presenting the competitive environment in a business plan.

VI.13 COMPETITIVE ANALYSIS

Our structure is better positioned for the market

Dimensions

Small or sale price formulas

Does the establishment offer low-cost packages?

Location
Is the establishment located in a dynamic catchment area?
Production method Is the micropower station capable of producing electrical energy?
Our structure has no competitors, we are in a healthy environment where there are no competitors.
You'll find this essential part, which explains how we can implement our strategy to be profitable.
We have adapted the strategic elements to suit the sector in question, and you (our partner) can also use these elements to find customers and increase our turnover.

VI.14 3-YEAR ACTION PLAN

The development of our micropower station and our entire range of services will take place in several stages.

YEAR 1

- Business plan
- Search for financial partners
- Work and improvements to the organisation
- Testing different electrical network distribution zones
- Creation of websites for suggestions and perception

YEAR 2

- Development and extension of electricity networks
- Massive investment to attract new customers
- Developing partnerships with local producers

You'll find a 3-year action plan, drawn up by our experts and tailored to the sector in question.
We have chosen to implement this plan for our company, and this part of the business plan shows that we are organised and have a long-term vision for our business project.

YEAR 3

- Market study with a view to possibly franchising the concept.
- Setting up a new provisional budget to open a second establishment.
- Hiring of a CEO to delegate tasks.

VI.16 BUSINESS MODEL CANVAS

This business plan focuses on the market opportunity, the project, the market, the strategy and the finance.
Here we look at how we can generate income and profit.

PARTNERS	KEY ACTIVITIES	

Equipment supplier Local authority Service providers Monitoring the electricity distribution network	Sale of electricity produced by the micropower station **KEY RESOURCES** Employees Electrical equipment Work tools	**VALUE PROPOSITION** • Authentic and local • Courtesy • Service

CUSTOMER RELATIONSHIP

Social networks

Face to face

Customer service

Website and blog

DISTRIBUTION

On site

Delivery

Electric transport

VI.17 MARKETING STRATEGY

Our strategy for attracting customers

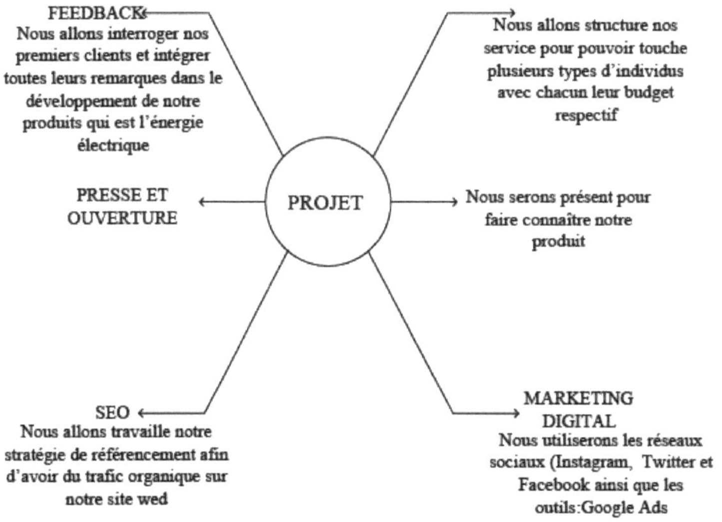

FEEDBACK

We will be interviewing our first customers and incorporating all their feedback into the development of our electric energy product.

PRESS AND OPENING

We're going to structure our services to be able to reach several types of individuals, each with their own budget

> We'll be there to promote our product

SEO <■

We are going to work on our SEO strategy in order to organic traffic to our our website wed

DIGITAL MARKETING We will be using social networks (Instagram, Twitter and Facebook) as well as the tools:Google Ads

VI.18 RISK MANAGEMENT

We take a proactive approach and have already identified potential risks and their management.

Table 14: Risk management

Event	Level of risk	Impact	Management
Delayed development	Medium	Very high	We have factored this risk into our plan to ensure that we meet our deadlines.
Gap between what we and this demand the market	Low	Very strong	We have already validated the interest of the market by talking to our potential customers and we are continuing to take feedback from users into account even though there is a micropower station.
Delays in the search for financing	Low	Very strong	Banks have shown an interest in us. We have funds available to start our project to implement of a micro-power station at KINZONO

This part shows that we are proactive, that we have considered the potential risks and have even already thought of a solution to mitigate the impact of these risks.

VI.19 A SOLID, PROFITABLE PROJECT

Six arguments that prove the economic and financial viability of our project.

WALK

We're in a buoyant market that's growing every year

NICHE

We operate in a niche where there is little competition

SOCIAL NETWORKS

We already have an active community following on the networks.

FINANCES

Our financial forecasts show that we will quickly reach our break-even point.

PARTNER

We are going to surround ourselves with partners who will support our growth.

EXPERIENCE

	The project leader has significant experience in 1 industry.

This final section (before the conclusions of the financial forecast) provides a convincing summary of the reasons why our project is viable and deserves funding.

VI.19.1 FINANCES

1. PROVISIONAL PROFIT AND LOSS ACCOUNT
2. PROVISIONAL BALANCE SHEET
3. FORECAST CASH FLOW BUDGET
4. PERFORMANCE INDICATORS
5. CALCULATION AND ANALYSIS OF BFTR
6. FINANCING PLAN
7. INITIAL INVESTMENT

Finally, the financial section enables the business reader to assess the future profitability and financial soundness of your business project.

We want our financial forecasts to build easily, these models are simple to use, complete and how to apply for operating funding from financial partners.

VI.19.2 PROVISIONAL INCOME STATEMENT

Over the next 3 years.

Table 15: Provisional income statement

Elements	Year 1	Year 2	Year 3
Sales figures	150 000$	200 000$	255 000$
Operating expenses	30 000$	35 000$	40 000$
Operating profit	31 000$	32 000$	33 000$
Financial income	30 000$	34 000$	38 000$
Finance costs	10 000$	12 000$	14 000$
Financial result	50 000$	51 000$	53 000$
Tax and import	10 000$	15 000$	18 000$
Result	311 000$	379 000$	451 000$

The projected income statement shows whether a company will make a profit or a loss over the coming years.

VI.19.3 PROVISIONAL BALANCE SHEET

Table 16: Forecast balance sheet

Elements	Year 1	Year 2	Year 3
ASSETS	80 000$	85 000$	90 000$

Non-current assets Current assets	50 000$	55 000$	60 000$
LIABILITIES Shareholders' equity Debts	85 000$ 90 000$	90 000$ 91 000$	93 000$ 94 000$

Over the next 3 years

The projected balance sheet is used to analyse a company's future financial position.

VI.19.4 FORECAST CASH FLOW BUDGET

Over the next 3 years

Table 17: Cash flow forecast

Elements	Year 1	Year 2	Year 3
Collections			
Operating cash flow	45 000$	50 000$	52 000$
Financing flows	60 000$	62 000$	69 000$
Investment flows	70 000$	75 000$	80 000$
Disbursements			
Operating cash flow	10 000$	9 000$	8 000$
Financing flows	8 000$	8 000$	7 000$
Investment flows	13 000$	10 000$	7 000$
Cash flow	10 000$	8 000$	7 000$
Cash balance	15 000$	10 000$	7 000$

Over the next 3 years

The forecast cash flow budget presents expenditure and forecasts in the form of cash flows.

VI.19.5 PERFORMANCE INDICATORS

Table 18: Performance indicators

Elements	Year 1	Year 2	Year 3
Sales figures	300 000$	350 000$	400 000$
Sales + actual production	100 000$	150 000$	200 000$
Overall margin	180 000$	185 000$	200 000$
Gross operating surplus	100 000$	105 000$	108 000$
Operating profit	90 000$	100 000$	105 000$
Current result	80 000$	90 000$	100 000$
Result	70 000$	80 000$	90 000$
Cash flow	80 000$	90 000$	100 000$

VI.19.6 WORKING CAPITAL REQUIREMENTS

Table 19: Working capital requirements

Elements	Year 1	Year 2	Year 3
NEEDS			
Stock of materials	100 000$	105 000$	110 000$
Trade receivables	90 000$	100 000$	160 000$
VAT receivables	100 000$	150 000$	160 000$
Resources			
Trade payables	170 000$	180 000$	200 000$
Social debts	160 000$	190 000$	210 000$
VAT debt	80 000$	85 000$	90 000$
Tax debt	58 000$	60 000$	65 000$
WCR	250 000$	300 000$	255 000$
Change in WCR	100 000$	105 000$	300 000$

Working capital requirements represent the funds needed by a company to finance its operating cycle.

VI.19.7 FINANCING PLAN

Over the next 3 years Table 20: Financing plan

Elements	Year 1	Year 2	Year 3
NEEDS			
Investments	400 000$	410 000$	500 000$
Change in WCR	300 000$	310 000$	360 000$
Bank loan repayment	50 000$	45 000$	30 000$
Resources			
Capital contribution	400 000$		
Current account contribution	300 000$		
associates	50 000$	20 000$	10 000$
Loan subscriptions	60 000$	70 000$	80 000$
Change in WCR	100 000$	110 000$	90 000$
Cash flow	10 000$	15 000$	18 000$
Change in cash and cash equivalents			

The financing plan ensures that a project is financially balanced by measuring its needs and resources.

GENERAL CONCLUSION

Here we are at the end of our study entitled : Analysis of the profitability of electrical energy produced by a hydraulic micro-power station, the case of KINZONO, given that energy contributes to the economic and social development of a nation, This is the reason why we wanted to provide the locality of KINZONO, which is in the commune of MALUKU, with a modern micro-power station enabling it to leave this rural stage for that of the city. Based on the study of objectively verifiable indicators, we confirm that it is profitable, given its necessity.

But more in-depth work on the community's expectations, assignees and development objectives is essential in order to generalise this micro - plant model.

The decision to set up a rural electrification structure must take into account not only the profitability of the project, but also its social impact and its effects on the environment.

Rural electrification is an important factor in the development of rural and urban-rural areas, but it is only one ingredient in this process. It needs to be integrated into a development programme with all the other ingredients in the process: education, awareness-raising, the environment (roads, etc.), health and social services, and access for the population to credit, inputs and other services.

In all cases, the right solutions must be a compromise between economic, environmental and social considerations, and above all involve the population as the main beneficiaries and players.

As energy is the driving force behind all socio-economic development, no-one is unaware of the significant contribution made by energy services to boosting all sectors of economic and social development, as well as their importance in all human activities for achieving better living conditions.

Despite the Democratic Republic of Congo's major assets in terms of energy resources (hydroelectricity), the overall electrification rate is too low, and even worse in rural areas (1%). Per capita consumption has been falling since 1990.

We suggest that the development programme be applied without fail, because that is where the problem lies in the Democratic Republic of Congo.

1. The country needs to develop and promote new and renewable energies, pending major development work on hydraulic sites, so that local electricity can be brought to households, even in rural areas.
2. Intensifying research and exploitation of sedimentary basins could reduce the Democratic Republic of Congo's overdependence on energy.
3. Faced with deteriorating terms of trade and an extroverted economy, we need to set up a genuine legal framework that is highly attractive to public and

private investment, while preserving the State's share, which must be significant.
4. Restore good governance through rigorous and rational management of financial and energy resources, taking into account the needs of the Congolese population.
5. Promote knowledge of operational and strategic management in the energy sector.

The Congolese state alone cannot respond effectively to these energy challenges because of the inadequacy of the government's financial resources, which have to deal with a number of priorities. The inadequacy of private initiatives in the energy sector and the lack of funding are some of the obstacles to the revival of the energy sector in the Democratic Republic of Congo.

BIBLIOGRAPHY

I. WORKS

1. APCE : (Agence pour la creation d'entreprises, creer ou reprendre une entreprise, 13e Edition, Edition d'organisation, Paris 2000.
2. BIZAGUET Armand, Small and medium-sized businesses. Que sais - je ? PUF, Paris 1991;
3. BIZARGUET. A, le secteur public et la privatisation, Paris, PUF, 1980, P52
4. BERNOUX; P, la sociologie des organisations Paris, Ed du Seuil, 2006, Pp 32 - 38.
5. BRUNS and STALKER, the management of innovations, Oxford, Ed oup. Oxford, SD
6. CHEVALIER, A. and ROLLET, J., Organisation industrielle, production, entretien et manutention, Librairie DELAGRAVE, Paris, France, 1975.
7. DE COSTER, Michel, Sociologie du travail et gestion du personnel, collection gestion et organisation des entreprises, Editions Labor, Brussels, 1987.
8. DESBASEILLE, Gerard, Exercices et problemes de recherche operationnelle, Dunod, Paris, 1976.
9. DIANKABU and DJIBU, critical study and proposal for the MT network.
10. DIEYE Alioune, Regime juridique des societes commerciales et du GIE dans l'espace OHADA, Acte uniforme sur le droit des societes commerciales et du GIE (AUSGIE) revise en 2014, 4th Edition, AZIZ DIEYE, 2014.
11. ENREGLE, Yves and THIETART, Raymond - Alain, Precis de direction et de gestion, Edition d'organisation, Paris, France, 1978.
12. Eric LUZOLO, Electrical Technology course note, 2^{ieme} Graduat ISPT - KIN, 2010 - 2011.
13. FAYOLE Alain, Entreprenariat, apprendre a entreprendre, 2eme Edition, DUNOD - strategie de l'entreprise, Paris, 2007.
14. FRANCIS PARIS, Missions strategiques de l'equipe dirigeante, Dunod - Entreprise, Bayeux, France, 1980.
15. Gode ATSWEL OKEL MUTONGI, Management approfondi, University of CEPROMAD/Doctoral School, Academic year 2020 - 2021.
16. HELFER and ORSONI, Marketing, vuilbert, 2007.
17. IBULA MWANA KATAKANGA, La consolidation du management publicau Zaire, PUZ, 1987.
18. JACQUES - DANIEL ROCHAT, Creer et gerer une entreprise, Editions ROC, Paris 2010.
19. KHUZAMA and VIKOIS, prepayment master 2010.
20. KHUZAMA and VIKOIS, si single - phase. Higt - distribution system 2010.

21. LA REPONSE AU DEFICIT du Leadership et de la Gouvernance en Afrique, Editions CEPAS, 2018.
22. LACRAMPE, Serge, Systeme d'information et structure des organisations, Ed. Hommes et techniques, Suresne, France, 1974.
23. LAFLAMME, Marcel, Dix approches pour humaniser et developper les organisations, Gaëtan Mortin et Associes, Quebec, 1976.
24. LAUZEL, P. coll. MUSSIER, G., Lexique de gestion. Entreprise moderne, d'Edition, Paris, 1970.
25. LE CONGO EN QUETE D'UN LEADERSHIP POLITIQUE de changement par des élections libres et democratiques. Conference - debate held from 2004 to 2007, Editions Cerdaf, 2017.
26. LEDEARSHIP AND RESPONSIBILITY, Association ISOKO - KIVU, 2010.
27. LUKENI LU NYIMI, Comment creer une PME au Zaire, formalites juridiques essentielles, 2nd Edition, Kinshasa, 1997.
28. MABI MULUMBA, l'entreprise privee installee au Kasai face a l'environnement economique national, Actes de colloque organisé[A] par l'Universite de Mbuji - Mayi.
29. MANUEL D'ORGANISATION, EO/FP, collection EO - formation permanente, Edition d'organisation, Paris, 1978.
30. MAXWELL, C. JOHN, Les regies d'or du leadership lemons apprises, Editions du Tresor cache, 2012.
31. MAXWELL, C., The principles of management, Priorite Education, January 2014.
32. MONTARETTO, MS., Manuel pour la direction du personnel, Ed. Hommes et techniques, 1972.
33. MUTSHI MUGUMO, Ferdinand, Les projets ; techniques d'évaluation et evaluation, Editions pensee Africaine 2005, 2010.
34. OKAMBAWA Wilfrid,. Le super leadership par le depouillement et l'amitie (Jn 3, 30), published by Lux Afrise, 2016.
35. OMOMBO OMANA ? Adrien, Le portefeuille de l'Etat et l'ajustement de l'économie de la Republique Democratique du Congo ; une autre maniere de comprendre la necessite de restructurer les entreprises appartenant a l'Etat Congolais, D/1997/Hipolyte Zere, Editeur.
36. REC/MODE, RURAL distribution and energy management in developing countries, March 2008, Perc electrification et transformalier, Kinshasa 2019.
37. Reform of 07 July concerning the transformation of public companies into commercial companies, public establishments or public services, public companies and public companies to be wound up.

38. REVUE POLITIQUE ET MANAGEMENT PUBLIC N°02, Vol 8, June 1990.
SNEL, Etude et suivi d'electrification transformaliere et rural, training Kinshasa, 05/07/2010.
39. TSHIBUABUA KAPY'A KALUBI, Benoit - Janvier, Cours d'action de direction et hierarchie fonctionnelle, taught at ENA - DRC, 2013.
40. TSHIBUABUA KAPY'A KALUBI, Benoit - Janvier, Cours de gestion previsionnelle des emplois, des effectifs et des competences dispense a l'ENA - RDC 2013, 2014 et 2015.
41. ZUKA MONDO UGONDA - LEMBA, Georges, Management et gestion de l'entreprise, Editions CEDESURK, Collection " Sciences sociales, Politique et Administrative " Iere Edition, Kinshasa 2012.

II. COURSE NOTES

1. Ingenieur Bernardin MBOL, Quality Assurance UNIC doctoral school, academic year 2020 - 2021.
2. Professor Angele NSAMAN: Seminar on the five functions of commandments UNIC doctoral school, academic year 2020 - 2021.
3. Prof Angel ONSIN N'SAMAN, leadership as a function of organisation and command, CEPROMAD, DEA, 2020. Pp 39 - 78.
4. Prof Bernardin M. Performance management, CEPROMAD, DEA, 2019.
5. Professor PALAMA Francois, rare scientific writing, UNIC 2020 - 2021 doctoral school.
6. REC/India, Rural power distribution and energy management in developing rural countries, March 2008.

III. OTHER DOCUMENTS

1. OECD, 2013, *Ten Principles to promote Better Management and Administration*.

IV. OFFICIAL DOCUMENTS

1. OHADA, Traite et actes uniformes commentes et annotes, 4eme Edition juriscope, 2012.
2. JOURNAL OFFICIEL, no. Special, 27 February 2013, various texts on the 2013 tax reforms.
3. OHADA, Addendum to revised and commented texts. OJ, February 2014, Juriscope, 2014.
4. JOURNAL OFFICIEL, no. Special, 28 March 2014, la reglementation du changer en RDC.

V. WEBOGRAPHY

1. BUGEIA.J.w.w.w.com, cours conducteur et cable, page consultee le 02/02/2021.wikipedia encyclopedie libre http// :

2. www. wikipedia.com.reseau electronique, page consulted on 05/09/2021.
3. http: //www.imf.org/extemal/np/sec/misc/qualifiers. htm
4. http://www2.ohchr.org/english/bodies/cedaw/docs/ngos/CCEDEF_DRC55_F orTheSession_en.pdf
5. GSMA, https: //www.gsma.com/mea/
6. http://w-28-ww.pointu-magazine.com/1
7. http: //www.pitou-magazine.com/1 -28 A VIVRE.php
8. http: //www.science. univ nantes.fr/sites/claudes sai ntblanquet/synophys/41 ene rgy/.htm
9.http://www.futurasciences.com/magazines/environnement/info/dico/d/enrgie_r enou_velablement-primaire_693
10. http://www.cea.fr/jeunes/mediatheque/animations-flash/energies/les-divers-sources-d-energie.
6. World Bank (2017) DRC Electricity Access & Services Expansion (EASE)

APPENDIX

Appendix 1

Management principles

1. Work specialisation

This principle depends on the size of the company. At a certain point, the company can concentrate certain tasks at a single workstation, but at another point, it has to set up specific tasks that should not be concentrated at a single workstation but spread across several workstations, where new bodies are created to replace the body originally responsible for these tasks.

2. Responsibility

It is the obligation to answer for one's own actions or those of others in the event of delegation of power or duties.

Responsibility is symbolised by the manager's authority or power to do or not to do something, which is justified by the manager's right to command and to be obeyed in a statutory or non-statutory manner (personal authority).

In view of the foregoing, liability may be of certain types: legal responsibility, which consists of a series of operations, either of taking legal action on behalf of the company or organisation as accuser, or of going to court as defender; economic responsibility, which is concerned with the results of the company or organisation, the success or failure of which is the responsibility of the person who has the authority or power of injunction and of being respected; Of these types, there is also the existence of socio-political responsibility, which consists of making the authority whose effects one is willing to accept for the actions taken, whether positive or negative, and finally, administrative responsibility is that which concerns the administrative actions of the authority and management where it has the power to do or have done and therefore the consequences may be immediate or remote.

3. Discipline

These are essentially obedience, diligence and submission, which must be respected in accordance with the agreements drawn up between the company and its employees. Authorities which do not respect the commitments they have made to their employees, and conversely, if authority is not respected by subordinates, are referred to as indiscipline, which should generally play a predominant role in the smooth running and normal functioning of the business or organisation.

It should be noted that in a company or organisation, discipline is a very important factor for the success of actions and for the advancement of the company or organisation.

In this respect, discipline is to the company or organisation the blood in the human body.

4. Subordination

This is the principle that allows us to observe that orders established between people or agents are respected so that some are dependent on others. This is the notion of dependence between the agents of a company.

Moreover, subordination is also a function of the dependence of the particular interest on the general interest.

5. The Command Unit

This principle requires that each agent must receive orders or instructions from the one above him.

To avoid anarchy, it is not advisable for a member of staff to receive orders from more than two line managers (dual command), otherwise they will find it difficult to apply the principle of subordination in the event of contradictions between line managers.

6. The management unit

It is about a single leader, a single programme for a set of operations aimed at the same goal. We must not aim for actions to be carried out by several departments at the same time, and we must not diversify, because we must observe the rule of coordination and convergence of efforts to achieve the company's objectives.

7. Centralisation or decentralisation

This is the principle that allows the company to bring together in a single centre of power all actions against those or operations that are attached to an authority in order to reduce the importance of subordination. It is a question of finding the limits between the company's agents by concentrating the tasks in a single head or authority, and on the other hand, decentralisation concerns the freedom to assign certain actions to others either by agreement of I

This can be done either by authorisation through an official notification or by statutory authorisation.

Whether it is concentration or decentralisation, it is recommended that the collaborative relationship be effective and that formal consensus be reached in the event of differences.

8. Hierarchisation

Within a company, this principle establishes a type of social organisation that specifies ascending power relationships between agents, where a whole is ordered by means of a relationship of order in which one agent is superior to another and the next.

Hierarchy is also a set of people who have the authority in a company or organisation to ensure that objectives are met. As a result, any action taken by a subordinate agent must be reported to the higher authority, which keeps and reviews the report.

9. Esprit de corps

Esprit de corps is similar to the unity of a company's staff. This is a strength for the company, because "there is strength in numbers", and we must strive to establish unity within the company so that when we work together we can only succeed and win. It is important to avoid splitting up, as dispersing forces often leads to failure. To this end, unity is a function of system thinking.

10. The order

It is the principle that advocates the reasoned and logical arrangement of things in relation to each other, with the aim of putting the company in a situation of regular use, tranquillity, harmony, discipline, accuracy, etc.

This order can be based on degree, rank, class, value, ... so that the company or organisation evolves in harmony where each place, each thing and each thing, its place must avoid anarchy if society has established its order. It's called social order, and it's a function of good organisation for better results (better performance).

It is a question of order in the structures, in the operation of these structures and in the missions or tasks for which each person is responsible.

11. Staff and job stability

This is the principle that requires the thing to return to its state of equilibrium after its duly established loss. Thus, when a member of staff has taken time to carry out the fixed duties of a given post, he or she still needs time to be reassigned in order to adapt to new duties and show aptitude.

It should be noted that stability of personnel and employment is not to be confused with routine, as it manifests itself in dynamism in solving business or organisational problems and in the ability to take appropriate action which excludes learning and useless stammering which has no sense of being on behalf of the organisation or the business.

12. Remuneration

Any action taken deserves a salary as a reward for services rendered or work done by an agent in a company.

This remuneration can take several forms: salary, payment by the day, pay, attendance fees, appointement, dividend, interest, profit, etc. However, it must also take into account several parameters that make it decent with the power to render consequent service to the employee.

It should be noted that this can be in cash, by the job, in pieces or in kind, and is not subject to the various deductions which curb its power and release in places which would encourage waste. It must be admitted that sociologically, remuneration must be the subject of responsible reflection when it is fixed.

13. Fairness

Is the principle of fair and equitable treatment, where everyone is treated according to their right or merit.

It should be noted that equity is a function of proportionality, which is often the subject of much controversy. E. Hay spoke of this in what we call the Hay method of responding to this principle for a company that wants to respond to its social responsibility.

Distributing the company's wealth equitably and fairly enables it to check the wage tension between employees vertically and horizontally, in order to avoid discrimination, which is a source of many industrial disputes.

14. The initiative

For any company manager who takes initiatives with his staff is infinitely superior to those who do not. Innovation and creativity must be the hallmarks of managers and company authorities, whether they are changing, modifying, adapting or integrating. Many authors have preferred to speak of principles as infinitives, in particular: planning, organising, selecting, directing, coordinating, commanding, reporting, evaluating, mobilising, controlling, developing, monitoring, encouraging, supervising, training, informing, communicating, diagnosing, analysing,... as regular actions that should be carried out by the agents in the company or organisation when it is functionally harmonious.

15. Keeping files and administrative documents in reserve.

For future and advisory use, near or far. The concept of correspondence is of paramount importance and requires the company or organisation to maintain their files and documents governing its administration. Proper safekeeping of administrative files and documents in a desired state of reserve is the principle administratively under capital supervision which generally the preference is to create a single and univocal cell for storage called the archiving cell which must depend on the principal authority because of its discretionary and confidential nature.

16. Recruitment to functions

Is the principle of testing the merit and performance of full-time work in better and acceptable conditions based on international standard norms.

It is desirable for recruitment to take place when there is a vacancy where competitiveness is recommended. For this to be objective, recruitment to vacant posts should be official and everyone should be free to take part if they meet the conditions.

It should also be noted here that recruitment can be carried out on a competitive or merit basis. Parachuting is a recruitment method often linked to various

functional problems in the future.
In addition, interpersonal relationships can influence recruitment, so this should not compromise the smooth running of the company or organisation when there is a change in the leaders operating at its head.
"You can't recommend someone if you don't trust them and if you don't share the same spirit, because the image that the recommended person reflects is that of the person who recommended him or her or of the person who recommended him or her".

17. The principle of change and adaptation to socio-economic and political contingencies.

This principle requires the company or organisation to have an operational dynamic in order to remain effective for any length of time.
Changing people deserves special attention if the company or organisation is not to be destabilised.
Structural reforms are conditional and must enable the survival of the company or organisation and must avoid complacency or the simple fact of tribal or racist satisfaction or for reasons of punishment or sanctions politically inflicted on agents.

18. The principle of self-regulation

Is the process whereby the company or organisation determines the conflict resolution mechanisms and the internal or external control mechanisms in order to establish the limits of authority.
Every employee has his or her rights and duties, and self-regulation makes it possible to prevent unrest taking hold within the company or organisation thanks to its strength in the application of the sanctions system.
This application is not to be confused with tolerance, forgiveness and the rejection of hierarchical appreciation, which can lead to injustice and the settling of scores.
It is often hoped that the principle of self-regulation will be linked to the policy of senior management, which must report to the higher authority in order to oversee its power and avoid subjectivity in the actions to be taken.

Appendix 2

The sixteen Nsamanist knowledges

nsamanist knowledge is :

1. Building a monument from scratch

The common thread that keeps the manager active and on the move has its roots in work. A Luba proverb puts this powerful idea into practice when it says: "a persevering man never fails to achieve a good result, because his ideas are a rich source of wealth".

Life is complex, even in organisations and businesses. It has its ups and downs, but the manager, imbued with this principle, will have to bring together all the resources at his or her disposal or look for them elsewhere with a view to fitting them into a rational combination for a positive result.

2. Braving fear

In management, fear has been identified as the enemy of people who believe in change or progress, a brake on emancipation. The story of David and Goliath in the Bible tells us all we need to know about this weakness, which is fear, because if David had underestimated himself, he would not have braved the odds to win the battle against Goliath.

In fact, the manager must brave fear, confront obstacles and danger with courage, and think in terms of the organisational and operational iceberg on his way, as a Yaka once said: "an antelope that comes out of the fire is neither afraid nor ashamed to flee despite the cries of the people".

Hence, those who do not defy fear cannot fulfil their potential, because courageous people are freed from their cradles and pass on their message, their contribution to the development of the company or organisation in which they find themselves, in order to survive and develop.

3. Knowing how to dare.

As our RAR model puts it so well, the manager or administrator of a new type of company or organisation remains a man of action. He knows when he can act, when he can take the right action to achieve a positive and essential result.

Not to dare is to remain lazy and idle, because the principle of trial and error is a particular form of knowing how to dare. Some managers have argued that the person who dares is imbued with the perseverance that motivates them to act. It should be noted that the manager is the one who, in a given environment, has to say what needs to be done to improve the effects imposed by the environment, because he needs to go beyond himself and use managerial strategies for better entrepreneurship.

4. Doing better-better

The new type of Manager is endowed with eternal creativity. He must turn fact into procedure, failure into success,

By seeking effectiveness, efficiency, rationality, relevance and performance in these actions, he must change and transform his environment. It should therefore be noted that the quality of the company or organisation depends to a large extent on the proportion of work done by the manager, because improving living conditions remains the manager's battleground, and more particularly that of the new type of manager known as the Manager-Leader.

5. **Taking calculated risks**

Considered as an entrepreneur par excellence, the Manager must take calculated risks.

When he analyses a situation and finds that the success rate of an action is 50^0 /0, he just jumps at the chance.

6. **Don't take on someone stronger than you.**

In other words, if you don't want to do what you don't believe in, you have to get round it or have it done on your side, because the great strategy for a manager is diplomacy and negotiation.

In fact, for the Nande people, it uses an acceptable management strategy, the common expression is that you have to break the pen, humiliate yourself with the Boss to learn what he does and then imitate him to become great in this field so that one day you become the Boss.

7. **Fighting poverty**

He must regard poverty as a vice because he himself is money. Someone clever was able to take advantage of Tubi's ugliness, a person who weighed 350 kg and ate 5 times a day, and he said to him: "If you want to continue eating well and earning money, I'll take you somewhere, I'll rent you a stall. Anyone who wants to see you will pay money. The publicity around your name will bring a game of fans. The proceeds from ticket sales will be divided up according to a distribution key. This is a strategy that will make this manager and Tubi rich. It's just one strategy among thousands.

8. **Working with entrepreneurs**

You have to know how to select and sort out the friends who can be useful in obtaining and achieving the objectives previously defined. Since it has been bitterly observed that poor friends impoverish the rich even more, because they can only live at the expense of the rich, it follows that they cannot help the Manager to achieve his objectives by their contributions.

In Luba culture, it is said that: "kunda ya bani itu boba nemata" which means "the beans of several people are prepared with the saliva of those involved" which means "there is strength in unity". However, each of the members of this unity must have strength because the sum of the different strengths and intelligences will constitute a great power.

That's why it's so important for managers to make an informed choice about the friends and colleagues who can contribute to the fulfilment and performance of their actions and activities.

9. **Knowing how to move forward in disorder instead of dreaming in place**

As a man of action and a taker of initiative, the manager and leader of a company cannot spend the whole day daydreaming or doing nothing. He will be in a hurry, because it would be better to undertake actions, even in a disoriented and scattered way, so that some of them lead to a positive result and others can be modified to achieve the same end.

As a Mbuun proverb says: "He who climbs a palm tree cannot look down". You have to keep moving forward to achieve your goal, listening to others, and initiating actions that contribute to your survival and that of the society in which you live, because it's better to make the wrong decision and have the risks calculated for you than to fail to make the right decision.

It's simply better to do that than to sit back and do nothing, because sooner or later improvements or corrections will only be made to what already exists.

10. Combining planning with improvisation

Planning is understood as a stage in the management process where the objectives to be achieved and the resources required to achieve them are decided, taking into account the environmental forces likely to influence the activity.

Any manager of the new type (Manager), after having planned an activity, may come across unforeseen circumstances in the course of carrying it out. In his nature, he cannot be emotional, he has to brave fear in the face of any situation and then for an unpredictable and imponderable situation, he has to give ad hoc solutions. The resolution of this unforeseen situation cannot make the manager feel embarrassed.

11. Instant managerial culture

As a refined person, with the culture of any organisation, develops aptitudes, reflexes and real competences for its good performance. He is therefore recommended to be a man with a spontaneous, instantaneous culture, not hesitating when faced with a case. He puts forward an appropriate response, i.e. to hunger, a meal; to thirst, a drink; to the need for money, work.

12. Have a model

A model is the representation of a complex system which is assumed to be simpler and which is also assumed to possess certain properties similar to those chosen. In other words, it is a model whose behaviour and ways of carrying out successful actions are to be emulated, and which is to be considered a Saint Patron. He is also obliged to be a role model for others, because he is a model, a way, a path, a method to follow in order to achieve one's goals.

13. Knowing how to change the procedure

If management is the science of changing facts into procedures. It goes without saying that the head of a company or organisation would be this activist who has to change facts into procedures in order to improve what the environment imposes on him, because it marks and conditions.

Without rushing, he has to analyse things because there is no point in running, you have to start at the right point, because he is equipped and competent par excellence, he cannot rush into making decisions or choosing initiatives without prior study.

14. Having a thinking head

At any given moment, all he has to do is think, create and initiate actions whose sole and most ardent wish is that his collaborators follow him and put into practice all the aspects he has targeted. Once he has identified so many actions to develop, he will leave it to his colleagues to put his ideas into practice. He must remain the one-man orchestra that touches on this and that.

15. Knowing that a lack of time is a waste of time

The Americans say, rightly or wrongly, that "time is money", because time should be given priority, and time is as much a source of resources as any other business or organisational resource, especially for transport companies. Anyone who is short of time and short of resources is someone who doesn't know how to use these resources rationally, because there are only 24 hours in a day, and that's true for everyone, so shortage of time is a way of not materialising negatively and not putting it to good use.

All transport companies are those that apply time resources wisely, because time is at the heart of transport activities.

16. Knowing when to intervene

When managing or directing a company or organisation, even if the manager is that person who touches everything, who creates, who takes initiatives, who has a role model and who is himself the role model for others, he must remain a wise and cautious man.

He has to be the one who knows how to intervene, at the right moment and at the opportune or desired moment, because you mustn't put off until tomorrow what you have to do today, because tomorrow doesn't belong to you and every time a problem arises, you have to intervene with appropriate solutions or solutions that lead to a positive result. Don't put off what you have to do until later.

A wise man once said: "Every moment requires an appropriate action".

Printed by Books on Demand GmbH, Norderstedt / Germany